Inside the Photon

Tony Fleming

Inside the Photon

A Journey to Health

Pan Stanford Publishing

Published by

Pan Stanford Publishing Pte. Ltd.
Penthouse Level, Suntec Tower 3
8 Temasek Boulevard
Singapore 038988

Email: editorial@panstanford.com
Web: www.panstanford.com

British Library Cataloguing-in-Publication Data
A catalogue record for this book is available from the British Library.

Inside the Photon: A Journey to Health

ISBN 978-981-4241-40-3 (Hardcover)
ISBN 978-981-4241-88-5 (eBook)

Printed in the USA

Dedicated to
my mother and father, Ronnie and Harry Fleming, who died recently both aged 93, my wife Dianne, and my two children Jane and Simon and his wife Jodie

And the Lord God formed man of the slime of the earth

—Douay-Rheims Genesis 2:7

Dedicated to
Roman and Christine Bauer
Michael, Tara and Michael Roman Colorio
Christopher, Jillian, Piper, Beckett and Ainsley Colorio

These are my parents, my children and my grandchildren, who have all supported my vision of that something . . . 'somewhere over the rainbow.'

Never regard study as a duty, but as the enviable opportunity to learn to know the liberating influence of beauty in the realm of the spirit for your own personal joy and to the profit of the community to which your later work belongs.

—Albert Einstein

Contents

Preface

Over the past decade several excellent, wonderfully detailed books on bioelectromagnetics (BEM,[a]) magnetobiology, biophotonics[b] and developmental biology have been published. These include compilations of efforts by large numbers of authors, each one an expert in his or her own specialised field. The result of these efforts has been to focus the community of bioeffects and medical researchers on the opportunities that exist at present in the field of medical therapy. We may be entering a 'golden age' of medical applications of BEM and biophotonics technologies. It seems centuries of effort are about to be rewarded. But we need to 'gang warily'[c] about where we are in time, and a historical perspective is required in order to appreciate our present position clearly.

Put succinctly the scientific method is needed to validate several recent theoretical and experimental discoveries. We now have a field theory that applies to cell division. This new theory is based on the mathematics of self-field theory (SFT), a new mathematical description of physics and biophysics that involves biogenic electric

---

[a] The abbreviation BEM is used for either bioelectromagnetic or bioelectromagnetics, depending on context; likewise the abbreviation CEM is either classical electromagnetic or classical electromagnetics. Similar comments apply to EM.

[b] Photons and biophotons are taken to be synonymous within self-field theory (SFT), where the term 'photon' arose from 'hard' physics and the term 'biophoton' from 'soft' biophysics. SFT sees the universe like a cell as a cradle of created life where cosmological and biological evolutionary processes come together in a unified physics. Those denying the reality of evolution on religious grounds should read Douay-Rheims Genesis 2:7, 'And the Lord God formed man of the slime of the earth', and the creation myths describing processes of evolution. Photons and biophotons may in fact differ in the same way a buckyball and an armadillo may both be described as 'balls'.

[c] 'Gang Warily' is the motto of the Drummond Clan of Scotland who supported Robert I (the Bruce) (1274–1329). The first author's mother was Veronica Agnes Drummond.

(E), magnetic (H) and acoustic[a] (A) fields. The theory is supported by several major lines of experimental endeavour conducted over the 20th century, including the work of Alexander Gurwitsch, Ross Adey, Raymond Royal Rife and more recently Fritz Alfred Popp. While existing biophysical theory is unable to give any detail of photonic effects, SFT with its new photonic level lends in principle support for all these experimental efforts. SFT also expands on the work of Charles Darwin, who 170 years ago initiated the theory of biological evolution.

In brief Gurwitsch in the former Soviet Russia performed a series of experiments on onion roots from before World War I (WWI) through to World War II (WWII) that revealed ultraviolet (UV) rays were involved in cell division and that there existed a particular frequency associated with cell division. In the 1930s Rife, who had earlier travelled to Germany, reported that by exposing patients to various frequencies of electromagnetic (EM) radiation medically beneficial results were possible. Adey, beginning in the mid-1970s, performed a series of experiments exposing embryonic chick cells to extremely-low-frequency (ELF), modulated ELF and millimetre radiation, which have been referred to as 'window'-type effects because of the way the results are seen at several frequencies and amplitudes around a central frequency and amplitude. Popp from the late 1970s has performed a large number of experiments using photomultipliers that have confirmed Gurwitsch's initial finding that photons of UV energies are emitted by strands of deoxyribonucleic acid (DNA). In all these cases the scientific and medical communities were ultimately unconvinced by the reported findings. Part of the ongoing problem seems to have been that in all cases other forms of medical and biological discoveries that were current in the period were seen as more promising and supported by mainstream biophysical theory. This includes the predominant place of atomic chemistry and the genetic theory of molecular biology. In all cases various scientific communities were unwilling to accept the findings due to a lack of a supporting theory and a feeling that the findings

[a] In a similar fashion to the way E and H fields are either via contact currents (amperic) or radiative (photonic) in nature, so too A fields are either vibrational (describing a range of macroscopic phenomena whose atomic quanta are photons) or radiative (phononic).

were incidental to the perceived underlying cause of the effect. Some experimenters reported failure to replicate results. But critically there was no theory with which to understand any effect or with which to compare the results. Now, however, there is SFT with which to perform the necessary validations by comparing experiment with theory.

A major link between SFT and Darwin's work is also part of our journey of discovery in this book. The evolutionary theory of life forms was first formulated by Darwin as he recorded his findings of the Galapagos Islands aboard the Royal Navy's *HMS Beagle* in the 1830s. Natural selection is a major part of this work, a key mechanism that involves the struggle for existence, competition, adaptation and fitness. These concepts have dominated evolutionary theory since Darwin's groundbreaking book *On the Origin of Species by Means of Natural Selection, or the Preservation of Favoured Races in the Struggle for Life* was printed in 1859.

In this book we examine how cells cooperate to achieve replication. There is seen to be a correspondence between cosmological evolution following the big bang and biological evolution. We investigate how cells cooperate within colonies to achieve adaptation of species to the slowly changing levels of EM and acoustic energy within the solar system and the galaxy as the universe expands. Tissues, including human tissues, can be classified via their relative permittivity, in particular the real part of the complex dielectric constant. This is a measure of the cooperation between cells as they seek to provide enough energy to achieve mitosis or meiosis. This may provide a most valuable quantitative measure of evolution over time. This understanding of cellular cooperation provides us with a new line of research to understand how tissues evolved. In addition we see how life adapts by using cooperation between life forms. This is in stark distinction to the competition of Darwin's theory. It appears evolution uses *both* competition *and* cooperation to achieve adaptation to changing circumstances, a broader theory of evolution than previously realised. Mutation is a concept involving blind chance and arbitrary happenstance—for instance, cosmic rays are often proposed to be involved in mutation of biological life forms. The evolution of tissues appears *not* to be a mutation but an adaptation to changing energy levels. Of course this is not to say chance plays no part in the universe or in biology. The overall result is a wider, deeper understanding of evolution and its relationship to cosmology. One major finding of this interaction between SFT, cosmological evolution and biological evolution is the possibility that life might be found throughout our own Milky Way galaxy and perhaps in other galaxies throughout the universe. Like the earth, biodiversity of life forms when analysed over cosmological time periods consists in total of a broad spectrum of photonic and phononic fields.

But our central subject matter, cells and their dynamics during cell cycles rather than their evolutionary response over cosmological periods, begins some 220 years ago with a truly amazing experimental discovery. Immediately prior to the French Revolution, Galvani[a] was dissecting a frog in his laboratory during a lightning storm. To his astonishment the frog's leg muscle twitched whenever his scissors touched its nerve. Again, during a follow-up experiment, the nerve twitched. On this occasion there was no electrical storm; rather an electrostatic generator was turned on in another part of the laboratory. This scientific breakthrough into the bioelectric basis of living tissues has taken 200 years to come to fruition. Galvanotherapy is an application of direct current that offers an alternative to the conventional treatment of cancers via surgery or chemotherapy. Galvani was cautious about immediately publishing his results, and he investigated further before publishing his *Commentarius*. He revealed that the legs of decapitated, skinned frogs could continue to move for considerable periods, as long as the necessary muscles and nerves were still viable. Importantly Galvani's experiments were able to be reproduced. The work was challenged by Volta, who did not accept Galvani's biophysical explanation. To some at that time, Galvani's experiments seemed like quackery, a trick of magic, while Volta saw the frog's leg as an electrical engine. Eventually Volta would invent the voltaic pile battery in an attempt to disprove Galvani's explanation of animal electricity. Volta was in fact greatly impressed by Galvani's work in physiology and recognised its historical importance.

---

[a]With regard to first or Christian names the authors have chosen in almost all cases to omit them for brevity. With regard references these are not cited in the context except for Chapters 4 and 5, where details of contextual citations are helpful.

Galvani and the wide scepticism towards his scientific results have been standard fare for those who have dared challenge the status quo of the scientific and medical fraternities over the past two centuries. Like Galvani there have been others who have incurred the incredulity of those who fail to see the glass half full and see only the glass half empty. The names might have changed, but the story has continued into the 21st century. It is the debate itself, as found in governments and institutions within the democratic world and beyond, that supplies the arguments and motivations for advancements, or otherwise, in all fields of endeavour. And this is our real aim in this present book—to supply the missing theory and join the dots found within the experiments that have been common knowledge now for at least the past century.

There are two main threads of bioeffects, one named BEM and the other named biophotonics. BEM in this text can historically be associated with classical electromagnetics (CEM) due to its macroscopic definition of bioeffects, while biophotonics can be associated with SFT with its understanding of a lower level of interaction between biological systems and EM fields. SFT is a recently discovered mathematics that describes a photonic level of EM interaction within atoms that also applies to molecules and in general across physics. SFT solves the Maxwell–Lorentz (ML) equations analytically in comparison to the numerical solutions of quantum methods. SFT uses modified systems of ML equations to study the weak and strong nuclear forces and the gravitational forces within the cosmos.[a] One major difference between SFT and quantum theory is the simplicity of the former compared to the complexity of the latter. No longer are pages full of spaghetti mathematics needed to understand the physics and biophysics. There is only one basic formulation associated with SFT, and this applies throughout physics and biophysics.

The current situation in relation to BEM and biophotonics is reminiscent of the investigations by Planck of blackbody radiation. Both Rayleigh and Jeans at long wavelengths and Wien at

[a]The term 'cosmos' indicates all matter and its space; if there are multiverses the cosmos includes all multiverses and the space they occupy en masse. On the other hand the universe indicates our particular universe as distinct from any other universe.

short wavelengths had experimentally obtained differing analytic equations for blackbody energy. The mathematics indicated an 'ultraviolet catastrophe' somewhere in the middle: CEM and the Rayleigh–Jeans law indicated an infinite energy response of the blackbody somewhere in the UV spectrum. Planck substituted a discrete series expansion for the exponential in both laws, showing the two classical laws and Planck's quantum expression agreed. Five years later Einstein published an experimental paper providing validation of the quantum nature of EM fields via the photoelectric effect whereby electrons are emitted by metals exposed to an incident monochromatic stream of photons.

This situation that occurred a century ago is also currently mirrored by the mathematics of theoretical physics. The tools of the theoretical physicist have been general relativity (GR) used to provide a mathematical framework for cosmology, and quantum field theory (QFT) that has been incorporated into the standard model of particle physics, the minute building blocks of matter. These tools strongly infer another mathematics that might unify the physics of the large and the small. One possible candidate has been the mathematical relationship between GR and QFT whose union is termed quantum gravity, although the attempts have been heuristic and of limited success. There thus remains a conflict between GR as a mathematical theory for the very large scale and quantum theory as a separate mathematics for the much smaller scale, similar to the Rayleigh–Jeans and Wien laws of blackbody radiation. In simple terms, no physical or mathematical rationale exists for the two different procedures, yet the mathematics of GR and QFT applies at present to the two ends of the size spectrum.

At the same time no single mathematical theory has served the bioeffects community. Neither GR nor QFT is suitable for predicting bioeffects. The best fit up to the present has in fact been CEM covering macroscopic effects down to but not including microscopic effects. While organs and tissues are macroscopic, cells and their myriad components are microscopic or smaller; ionic channels in the plasma membranes are nanoscopic. Although Maxwell and many others since have used CEM to model cells, including the use of lumped circuit analysis, such models are unable to accurately model the dynamics of cells, their ionic currents and their biogenic fields. In other words the biological system interacts between its constituents. Capacitive coupling means the analysis must take into account all parts of the system.[a] There is reason to believe inductance is overlooked by microdosimetric techniques that seek to classify three-dimensional regions via two-dimensional techniques; yet magnetic effects may be crucial in understanding cellular dynamics as we shall investigate. The fact is there has been no generally effective mathematics at the microscopic level or below. The overwhelming result of the application of SFT across physics and now biophysics is the finding that there has been a lack of magnetic theory resulting from an incomplete mathematics associated with quantum theory. Further compounding the problem, QFT has other serious theoretical limitations. Uncertainty and renormalisation are two areas where numerical difficulties arise. Another theoretical problem is the heuristic basis of all quantum methods where Planck's constant, obtained from experimental evidence, is inserted into a wave-like Lagrangian equation assumed to hold true at the atomic level. Atomic chemistry with its probabilistic orbitals is the result of quantum mechanical computations. While there have been many successes of atomic-level chemistry, intrinsic theoretical hurdles in its mathematics do not allow a biophotonic-level analysis.

Thus at the present time, many reported bioeffects cannot be validated via the scientific method without a mathematical theory to support experimental findings below the macroscopic level. Before SFT there was no theory to support the experimental findings that very small flows of photons of the order of a few hundreds, perhaps even less, could eventually cascade into significant biological effects.

However, we are finally in the fortunate position of having a recently discovered mathematical theory to support the experimental findings concerning ultra-weak energies associated with

[a]Mirroring the way capacitance couples across the entire system, Einstein via the Einstein–Podolsky–Rosen (EPR) paper published in 1935 tried to demonstrate the incompleteness of quantum theory to no avail. It is only in the light of SFT that the mathematical reason for the incompleteness of quantum theory is understood. The mathematics of SFT is intimately linked to Heisenberg's uncertainty principle (HUP), where the equations at the heart of SFT are seen to be almost identical to those of HUP, except that the inequality is replaced by a single equality and one equation becomes two to accommodate both electric and magnetic currents associated with a charged particle.

these tiny photon flows. Over the past 200 years the mathematical theory of E and H fields has developed to the point where we finally understand how the binding energy of atoms and molecules works. SFT also gives us an understanding of the phonon field, the quantum of acoustic energy and its relation to high-energy photons within nuclear physics and the production of the gluon. With regard photons of EM energies, they play a crucial role in atoms transiting back and forth between atomic particles to bind atoms and molecules together. SFT has extended knowledge of EM so that it is now known how E and H fields work at the atomic and molecular levels. This is a theoretical breakthrough beyond quantum theory with its uncertainty. The photon can be treated analytically as a composite system composed of sub-photonic particles. Theoretically we can surmise how the boson family exists at differing energies. This gives rise to '*photon chemistry*' whereby the atomic level chemistry has a new formerly unknown level of organisation that gives information not only about bond lengths and angles but also about dynamic speeds of rotation that tells us exactly where the atomic particles are at any time. Like the hydrogen atom the photon has radial and spin states and can exist as compounds such as the gluon within the strong nuclear region. This view inside the photon reveals a completely new perspective of EM far beyond classical EM and the quantum theories of the 20th century.[a]

Knowledge of bioeffects due to E and H fields has in fact mirrored our understanding of the mathematics of the fields. Within a few years of Galvani's findings, Coulomb discovered the inverse square law of electrostatic fields. Some 30 or so years later Ampere discovered a similar equation governing magnetostatic fields. In 1864, Maxwell formulated 20 quaternion equations, which

[a]A companion text to this present text gives an outline of SFT: *Self-Field Theory: A New Mathematical Description of Physics*, Pan Stanford Publishing, 2011. 'Following the recent developments that have evolved from the classical electromagnetics of the electron's self-fields, this unique perspective introduces self-field theory (SFT) as a new mathematical description of physics distinct from quantum field theory (QFT)—the physical theory of choice by physicists at the present time. This informative report is the foremost study of how SFT is capable of obtaining [a] closed-form solution for all atomic structures—rather than the probabilistic solutions of QFT—due to its bi-spinorial motions for particles and fields that obviate uncertainty.'

Lorentz and Heaviside transformed into the four vector equations named in Maxwell's honour. These four equations together with the macroscopic description of the E and H fields are termed classical EM. Failing to find a solution for the atom at the turn of the 20th century, CEM was seen to be inconsistent at atomic domains and smaller. Lorentz and Abraham had used CEM to investigate the self-fields of electrons without success. Planck's investigation of blackbody radiation and Einstein's photoelectric effect indicated a new phenomenon that heralded quantum physics. In 1927 quantum theory replaced CEM as the mathematics of the atom. This led to chemistry as we know it today, where quantum mechanics (QM) can predict bond lengths and angles. But QM tells us very little about the binding energy, in particular how photons operate at this level within the atom. That was left till 2005, when a modern version of the SFT of atomic E and H fields and how they bind electrons to protons was uncovered by examining mutual effects within atoms.

SFT was successfully applied to the hydrogen atom. The breakthrough came when instead of looking at a single particle, the electron of Lorentz and Abraham pairs of particles, the electron *and* the proton were studied. If the founding fathers of classical EM, Maxwell and Lorentz, had been told that SFT could solve the vector equations named in their honour as closed-form analytic expressions, they would perhaps not have been surprised, as they learned their mathematics before the advent of quantum theory. To them analytic closed-form solutions would serve to verify the physical basis of EM, nothing more nothing less. On the other hand if Heisenberg had been told the same, he would have been very surprised, since to him uncertainty was a part of the fabric of reality and his quantum world was indeed uncertain and never knowable beyond a level of uncertainty. If Einstein had been so informed he would have been delighted as he was unconvinced of the probabilistic nature of quantum theory. A unifying principle beyond uncertainty that enabled physics to be described in a deterministic fashion eluded him during his later years. That is precisely what SFT is, a unifying mathematics that applies across the breadth and depth of our scientific knowledge, mathematics, EM, photonics, acoustics, physics, chemistry, nuclear physics, biology, astrophysics, cosmology and geophysics, to name a handful.

SFT's success has been to solve the ML equations for the hydrogen atom using bi-spinors, particles and fields moving with double rotations. What this Bohr-like solution reveals is a previously unknown photonic layer of solution within the hydrogen atom that explodes our present knowledge of physics and chemistry. Now a more detailed description is available of the physics gained from quantum theory over the past 80 years. In the same way that Planck's quantum series played a vital role in solving the blackbody radiation problem, SFT provides a single mathematical formulation that solves at both large and small domains of GR and QFT. Hence the analogy to Planck's solution to the blackbody problem is now repeated to take us 'beyond quantum'.

In essence this text follows the history of E and H fields over the history, indeed prehistory, of the scientific record of mankind, in particular over the past two centuries. Scientific knowledge is much older than moderns appreciate as traditional knowledge of the constellations and 'bush medicine' of Australia's indigenous peoples attest. This is mirrored by an evolving understanding of the effects of E and H fields upon biological systems. The early Chinese and North American Indians used the local geology to select where to live, grow seeds, farm, fish and hunt and in what orientation.

It is a coincidence of modern science and development that Galvani performed his revolutionary biological experiments, Coulomb discovered the mathematical formula for electrostatic E fields and the Industrial Revolution occurred within a relatively short time frame towards the end of the 18th century. In retrospect we see the origins of the modern world with its benefits and problems and how this contrasts with the early understanding of the aboriginal peoples of the world and in general how they carefully husbanded the bounty of the earth surrounding them.

The history of man's development since the Industrial Revolution has depended, in large part, on the techniques learnt from scientific endeavours. Adam Smith's economic rationale had proposed a future full of promise, but by the mid-19th century this ideal had devolved into sweathouses where worker's lives and health were scant-considered in comparison to the dominating profit incentive. The situation was out of balance. A balance was restored by the introduction of unions designed to encourage workers to act in unison. Since that time, the wheels of industry have turned reasonably well except for periods where the strains within the global economy have caused periodic meltdown of the kind we are presently enduring. The situation is again out of balance. This time there is an impact external to the economic world that is becoming visible: there is an impact of doing business as usual in the form of what is termed '*global warming*'. In particular the continued use of fossil fuels is polluting our atmosphere and oceans, adding to the likelihood of a runaway greenhouse effect. With the emerging economies of China and India entering the global economy the situation may get worse before it can get better. Science may need to rethink its methods and provide cleaner ways of doing business.

What appears to be happening is that there is a gap between knowledge as applied to industry and the actual underlying scientific reality. In our efforts to live in the modern world with its ethos of production and consumption, our industrial methods are falling short of maintaining a balance between the economy and the world we live in. Most pertinent to this text, the same appears true of our medical health within the context of scientific knowledge over the course of the last century. The use of pharmaceuticals since the implementation of QM in the 1920s and its impact upon atomic chemistry can be seen to mirror the effects of industrialisation on the modern world. Atomic chemistry as it applies to both industry at large and the pharmaceutical industry in particular has aroused a level of retrospective suspicion as to its ubiquitous side effects, reflecting a lack of knowledge about the underlying scientific reality. With regard to HUP, is there a significant effect due to what we don't know of what we are not taking account? The facts appear to speak for themselves: there is an imbalance at work, a lack of sustainability. Like the financial cycle we may mitigate our situation by taking account of long-term trends rather than focus primarily or even exclusively on short-term profit-maximising goals.

Our aim in this text is to describe a newly discovered mathematics that applies to the whole of known biology. It is a formulation about dynamic balances that apply to forces, including those due to E and H and A fields. We find reference within ancient Eastern medical practice to a balance within the body in terms of the yin and the yang, two complementary opposites. In overall thrust this sounds very much like SFT, and it is here we begin our medical journey.

In Chapter 1 an introductory history is given along with a range of issues involved in what in human terms is the most important scientific subject of all—medicine. Put bluntly, science always finds itself between a rock and a hard place, not wanting to be sold snake oil but trying to avoid throwing the baby out with the bathwater—the not insignificant task of walking on top of the fence. To the authors there appear certain effects that have been cast aside as pseudoscience by the prevailing wisdoms of the past. In fact these effects may have been scientifically valid all along. In retrospect the examples mentioned in Chapter 1 may have a scientific basis. Without being presumptuous, the work of validation is yet to be performed, but the indications at this stage are that such validation may well eventuate in a number of these cases.

In Chapter 2 the history and development of the mathematics of electric and magnetic fields is outlined. This history can be analysed into four sets of equations: electrostatic and magnetostatic field theory, CEM field theory, QFT and, finally, since 2005, SFT. SFT depends on its bi-spinorial description of both the EM fields and the interacting particles. Providing an analytic estimate for the mass of the photon, it also provides a simple hydrogenic structure within the photon. Its eigenstructure involves arrays of molecules of varying photon states that range in molecular flexibility from solid, liquid crystal, liquid and gas phases. Its structure provides an organisation reminiscent of the atomic chemical table where the phonon, the photon and the gluon can be related. SFT obtains an analytic expression for Planck's number, providing a basis for its understanding as a variable of motion applying equally to the electron, the proton and the photon. The fields of SFT differ markedly from those of CEM and QFT.

In Chapter 3 the connection between biological and cosmological evolution is discussed. It appears biological evolution depends on cosmological evolution. Life could be widespread within our galaxy. The various species form a 'tree of life' that specifies where each fits in terms of the time line of the earth's evolutionary history, beginning with algae some billions of years ago down to present-day mammals. The gravitational structure of our galaxy obtained via SFT may explain a recently discovered biodiversity cycle of 62 million years. The analytic form of gravitation is fundamentally different to Newtonian gravity and Einstein's GR. The SFT form of gravity at the galactic level incorporates the phonon, photon and gluon into a strong nuclear interaction.

In Chapter 4 the bioeffects due to diffusion or ionic currents within biological systems are examined. Biological diffusion consists of random thermal motions plus the coherent E and H field offsets of charged and dipolar entities. SFT brings a number of fundamental factors to the theory of diffusion. Biophotons move in helical motions between charged particles. Having an internal structure biophotons can change state suddenly across the spectrum as the energy density changes, inducing cascades at discrete energy levels or photonic frequencies. Biophotons also modify the energy density of a region. Chemical cascades occur at both thermal and non-thermal energy levels Cascades and other non-thermal effects are omitted from CEM heat analyses. Planck's blackbody theory applies to thermal and non-thermal frequencies, for example, rotational effects are ignored. Rotation of cell nuclei or proteins within cell membranes may help explain windowed observations of Adey at low levels of ELF and modulated ELF signals.

In Chapter 5 the discussion turns to cell replication and an SFT process the result of cell–cell interactions of both bioelectric and biomagnetic fields arising from the diffusion of dipolar proteins within the plasma membranes of a colony of cells. Two biophotonic feedback loops can occur in a colony of cells, (1) electrical feedback between membrane proteins of a fertilised cell and those of surrounding cells and (2) magnetic feedback between biophotons emitted by chromosomes and proteins diffusing within the plasma membranes of neighbouring cells. Experimental observations of metaphase and dielectric theory support the hypothesis, including Gurwitsch's pioneering work with onion root growth in the 1920s–1930s where growth tips can be modelled as a constrained membrane diffusion process.

In Chapter 6 two therapeutic methods are examined. Medical science has recently embraced the general modality of using non-ionising energies of static, EM, ultrasonic and audible A fields as a means of delivering therapies. In the first method non-thermal levels of pulsed electromagnetic fields (PEMFs) induce significantly enhanced concentration levels of Ca^{2+} acting as a second messenger on to the surface of cells in the chemical pathway associated with regeneration. The use of non-sinusoidal waveforms, including pulsed waveforms, creates time differential components of Maxwell's equations that are larger than the sinusoidal case. In the second case a spatially variable H field is used in conjunction with a frequency variable A field. This method uses the more penetrative powers of A fields relative to EM fields within the tissues of the body. The fields are applied to the body surface via an applicator that incorporates a permanent magnet. The applicator vibrates, including the magnetic head, at a frequency under therapist control.

In regard to the writing of this book, the principal author wrote the majority of the text in order to achieve cohesion across chapters, while the second author provided much needed assistance, especially with regard to the second method of therapy discussed in Chapter 6, the acoustic and magnetic applicators. The results obtained were solely her work,[a] including the frequencies and why they were used in all therapies using the CYMA1000®. The chapters on diffusion were developed from the principal author's own PhD, parts of which have been incorporated into the text. This includes the numerical diffusion work in Chapters 4 and 5. The constrained theory of tip growth was developed with Dr Rosemary White, currently of the Commonwealth Scientific and Industrial Research Organisation (CSIRO), who was, around 1995–1996, at Monash University's Department of Evolution, Growth and Development. At the time, for various reasons this diffusion work was unable to be published at that time, although it was prepared as two joint papers with Dr White. But that was to the eventual benefit of the work, which was much broader than initially considered. It was some years later, around 2003, when he, Elizabeth Bauer and Anthony Fuccioni spent two to three years involved in a three-way discussion about how the cell might achieve replication in terms of the E and H fields. Bauer suggested the hydrogen bond was important inside and outside the cell, Fuccioni suggested the

[a]For a readable guide to the method of therapy used by Elizabeth Bauer, the reader should read Gibbs C., *Cymatherapy—A Practical Guide for Everyone*, Paragon Publishing, Rothersdthorpe, 2010.

chromosomes act somewhat like a magnet at metaphase, while Fleming suggested a neighbourhood of cells acted cooperatively to induce a field enhancement on a central fertilised cell. Hence the overall process presented in Chapter 5 began as a joint effort and finished as a mathematical theory of cellular replication where the E and H fields are tied together within SFT with its photonic structure and collisional view of atomic and molecular binding energy and its photonic mechanics of molecular cascades.

Some of the appendices are based on the first author's papers published by the various symposia known collectively as Progress in Electromagnetics Research Symposium. There is also a frequently asked questions (FAQs) section, which sets out the various questions that have arisen.

Finally the authors are indebted to the editorial staff of Pan Stanford Publishing for their assistance in the production of this book, especially Mr Stanford Chong for suggesting the publication of both books and Mr Sarabjeet Garcha for his tireless editorial assistance. They would also like to give credit to the numerous illustrators whose work has been donated to the repository of Wikimedia Commons, in general a public domain collection of various media maintained by Wikipedia, the Free Encyclopedia. Wherever the term (credit: wikipedia) is given in a figure caption, the diagram, drawing, or picture has come from Wikimedia Commons, for which the author is very grateful. There are others unmentioned who have contributed to the work; as Sir Isaac Newton famously commented, 'All science is a process of seeing further by climbing on the shoulders of those previous scientists whose efforts have accumulated as the tome of work that represent an emerging, more accurate scientific truth that allows humanity to live a better, healthier life as time marches forward.'

Tony Fleming

Chapter 1

Introduction

Over the long history of humankind, various techniques have been tried to mitigate the effects of various illnesses and diseases. In ancient times it was widely thought good health and harvests, and by some life itself, came from the sun and stars. Generally medicinal therapies were painstakingly built up at the communal level, for example, by trying plants and other flora—what worked and what did not. Two ancient examples are herbal medicine and acupuncture, while homoeopathy is a more recent method. As knowledge of chemistry and biology grew, traditional therapies were shelved, while the East and indigenous populations retained them. No basis was found for many methods by modern science. In 1927 quantum theory was widely adopted across science, including chemistry and biology. Einstein, the most eminent scientist of the 20th century, considered quantum theory incomplete, along with its probabilities. But no other solution was available in this period when quantum theory dominated. Now, a new description of physics and biophysics, the mathematics of self-field theory (SFT), suggests re-examination of many alternative therapies since SFT shows quantum theory is indeed incomplete. In our modern era there are concerns and opportunities with electromagnetic (EM) exposures. Technological exposures can be many times natural

Inside the Photon: A Journey to Health
Tony Fleming

ISBN 978-981-4241-40-3 (Hardcover), 978-981-4241-88-5 (eBook)
www.panstanford.com

levels, while 'frequency medicine' involves exposures to EM or acoustic fields. Major advances have emerged since 1927, including physiological, immunological and genetic understanding and how human life develops from a macroscopic embryo; but chemistry is incomplete without the photonic layer of structure. Quantum theory has inadvertently masked the perspective of this unknown layer of photonic structure beneath atomic chemistry. Using this new mathematical physics we review recent scientific history. This knowledge refocuses us on the sun, the galaxy and cosmology. SFT adds much to our understanding of biology at the electric (E) and magnetic (H) field level, including cell division. Other possible areas of application of SFT, such as intelligence, memory, cognition and emotional health, currently remain scientific mysteries.

1.1 Introduction

At the turn of the 20th century a number of important scientific discoveries were made concerning light. These findings would set the scene for the century ahead. The most mysterious of these, without doubt, was Röntgen's discovery of X-ray photography in 1895. To the classical 19th-century physicist this effect was greeted with incredulity. Röntgen himself was reluctant to reveal the accidental finding until he had a chance to investigate further. In 1900 Planck discovered that blackbody radiation could only be explained if the photon frequencies in the equations he was studying were taken to be discrete. This supported the spectroscopic findings of the hydrogen atom in the late 1800s. Einstein's work in revealing the photoelectric effect in 1905 had validated Planck's assertions about a quantum effect. Einstein then rocked the foundations of physics by the amazing knowledge contained in his theories of special and general relativity. The idea that EM energy came in discrete relativistic packets necessitated a rethink of classical physics. The concept of 'field' had been developed by Faraday and Maxwell in the mid-19th century, but now the wave model of light was challenged by this particle behaviour. Was the photon a particle or a wave? What was relativity, and why did Maxwell's equations obey relativistic transformations? Hertz had demonstrated the feasibility and later Marconi the utility of radio waves. Röntgen's X-rays were able to provide medical imaging of tuberculosis, but the science of the photon at the heart of this high-energy application was primitive and raw. The photon remained scientifically misunderstood during the 20th century. This book concerns recent answers to the various questions about the photon that arose early in the 20th century. What must be understood is that this new view of the photon is based on the firm ground of exact solutions to the Maxwell–Lorentz (ML) equations and their new description of physics and biophysics. Prior to SFT exact solutions to the ML equations controlling atoms were unknown, and this lack was the driving force behind the formulation of the numerical solutions of quantum theory formulated in 1927.

The Great War of 1914–1918 was followed by a period of metamorphosis in human affairs, generally including the rise of national socialism in Germany and the Wall Street crash. Science and scientists were caught up in this whirlpool of world change that continues today. In the build-up to World War II (WWII) there were a number of important discoveries relating to the photon and the biophoton that occurred within a few years of each other from the 1920s into the 1930s: (1) Hubble discovered galaxies outside our own Milky Way in 1923; (2) Gurwitsch discovered what he termed 'mitogenetic radiation' when studying onions and yeast also in 1923; (3) from 1921 to 1934 Royal Raymond Rife invented a 'dark field' microscope, investigated resonance effects upon microbes and opened a medical clinic that was eventually closed down by US government authorities; (4) the electron microscope was invented in 1931 by Ruska and Knoll, breaking through a fundamental limit on microscopy using ordinary light, enabling study inside cells, including the cell nucleus; and (5) quantum mechanics (QM) was formulated in 1927 by Dirac, bringing together the wave mechanics of Shrödinger and the matrix mechanics of Heisenberg. Being a mathematical formulation this quickly filtered across many scientific disciplines, where it remains today.

There is a moral metaphor contained in the overall history of all these discoveries. During WWII the all-dominating high-energy application to atomic weapons was deemed necessary to end the war at the expense of the quiet, much more subtle effects of cell replication and embryology. This domination carried across the intervening years till recently. Thus a quickly developing scientific process of chaos, war and death were opposed to another slowly evolving process of peace, tranquillity and life. The undiscovered latent knowledge of biology and medicine was effectively stifled by the mathematics of quantum theory that dominated across science until the 21st century. Quantum theory has recently been shown to be incomplete, as Einstein suspected and tried to demonstrate in 1935. Although at first amazingly successful in some areas, including molecular biology and genetics, quantum theory has, obviously without intent, stood as an impediment to advances in areas of physics, chemistry and biology. Perhaps the world was not ready for the developments in understanding cell division . . . until now. Now there is a new layer of interaction. The photon acts as a binding agent not only between atomic particles but also within molecules that can interact to a stronger or weaker degree. There is a new layer of molecular and gravitational information contained within this interaction. Science has advanced beyond its previous limit to add a new photonic layer within biology, pharmacology and medicine.

Hubble used the 100-inch telescope at Mount Wilson Observatory to determine that the local group 'stars' were in fact other galaxies outside our own Milky Way galaxy. His observations were made in 1922–1923. Hubble went on to discover a cosmological red shift that played a major role in the theory of the big bang, the origin point in space-time of the universe. This discovery was a direct outcome of improved experimental photonics applied at the cosmological end of the size spectrum. At the time the view within classical cosmology was that the known universe was nebula-like, while the term 'galaxy' was not used in technical papers. The discovery that our Milky Way galaxy was one of many galaxies was mind-expanding for the classical mind-set of the early 20th century. As far-reaching as his work was, Hubble could not have foreseen the role galactic dynamics or the big bang, both resulting from his work, would play in the study of the biophoton, the quantum of life, its evolving energy and the determinant of all biological life forms.

As Hubble was discovering galaxies, evolutionary biology was deeply involved in a debate concerning natural selection and the struggle for survival proposed by Darwin in his classic text *On*

Figure 1.1 Sculpture of Mendeleev in Saint Petersburg (credit: Wikimedia).

the Origin of Species, published in 1858. Around this same period in the mid-19th century Mendel revealed the basic mechanisms involved in genetic inheritance, while Mendeleev discovered the periodic table of the chemical elements (Fig. 1.1). Since then the development of better microscopes and the breeding of short-lived species, such as fruit flies and bacteria, for research have led to a growing understanding of replication, development and the cell cycle. This includes the realisation that other factors apart from natural selection are involved in biological evolution, and this expanded view has become known as the synthetic theory of biological evolution.

Figure 1.2 The 100-inch telescope used by Hubble at Mount Wilson.

Cosmological knowledge increased significantly over the course of the 20th century. As well as Hubble's finding of the red shift and its relation to the theory of the big bang, his pioneering work in revealing a universe full of galaxies (Fig. 1.2) has eventually led to the dynamics of galaxies being studied at the start of the 21st century with respect to biological evolution on the earth. Galactic dynamics turns out to be a key to understanding evolution and biodiversity. Definite cycles of a period of 62 million years have been discovered in the marine fossil record corresponding to the galactic tides. Biological evolution has been put on a modern footing along with its cosmological counterpart, finally emerging from its classical origins.

Already in the early 21st century, along with the recent finding of biodiversity cycles, a new mathematical physics has been discovered relating to Einstein's general relativity discovered in 1915 and the quantum theory discovered in 1927. This new description of physics and biophysics evolved out of the classical electromagnetics (CEM) of the electron self-fields studied by Abraham and Lorentz in 1903–1904. Due to its bi-spinorial motions for particles and fields that obviate mathematical uncertainty, SFT can obtain closed-form solutions for all atomic structures rather than the probabilistic solutions of quantum theory. SFT unifies the known physical forces of electromagnetics, strong and weak nuclear forces, gravity and acoustics under a single theoretical umbrella, providing insight into the structure and origin of deoxyribonucleic acid (DNA) as a piecewise bi-spinor. Different analytic forms of gravitation within the universe, for example, a tri-spinorial galactic gravitation, suggest cosmological evolution and biological evolution are related.

The wavelengths associated with DNA can be compared to the EM emissions from the sun. Similarly the sizes of the evolved species can be compared with the magnitude of the sun's EM emissions. The sizes of the various life forms, from single-celled organisms to large mammals, relate to the energy available on the earth as a function of time elapsed since the big bang and the process of universal expansion. SFT sees biological evolution as a resonance phenomenon related to cosmological evolution. These resonances may apply at each domain of gravity to form a specific fingerprint or bar code of any evolved species. Biological evolution is thus not a random process when viewed at the cosmological perspective.

From the terrestrial domain Darwin saw struggle, randomness and uncertainty as the primary drivers of evolution. However, resonance appears involved at the cosmological level and the various gravitational domains below this universal level. As a result of the uncertainty within quantum theory it was considered there was a connection between quantum theory and cosmology. In 2006 Hawking and Hertog claimed QM forbids the universe from having a single history (http://www.sententdevelopments.com/2006/12/quantum-mechanics-forbids-single.html). As long ago as

1930 Fisher proposed a relationship between QM and biological evolution. But in the same way that SFT sees quantum theory as incomplete these relationships give an inaccurate view of both the big bang and biological evolution. The exact solutions available via SFT thus impact the opinion of both 19th-century Darwinian evolutionary theory and the 20th-century idea that quantum theory was the 'fabric of reality' and the way biological and cosmological evolution worked at a mechanistic level. Heisenberg's opinion was that the internal state of the photon could not be completely known; only a probabilistic solution of atoms and cosmology was available. Einstein felt that this view was incomplete, as did Shrödinger with his cat. Quantum theory was indeed incapable of seeing the photonic and biophotonic layer of interaction within atoms and molecules. But that does not alter the fact that this layer of interaction exists and can be explained deterministically via SFT. What is surprising is that we find we must understand cosmology with its various gravitational domains before we can appreciate what makes up our own bodies and that of our fellow earthling species. Yet to do this we must also learn the mathematics of SFT to appreciate how species replicate at the biophotonic level, the *lingua franca*, the *coin of trade* across the cosmos. Note as we proceed the way this biophotonic view blends into the classical chemical view, so they merge into a single perspective.

1.2 Introduction to Self-Field Theory and Biophotonics

Life in the industrialised world has given modern man devices and opportunities that were unimaginable two centuries ago. However, we are now bathed in a variety of E and H fields that are often much higher than the field levels that existed before modern times. The local geomagnetic flux density is around 1×10^{-4} T (1 gauss), depending upon the position on the globe. Some medical devices, for example, magnetic resonance imaging, involve exposure to fields above 1 T. Environmental E field levels vary between 150 V m^{-1} in fair weather and 10 k V m^{-1} during thunderstorms. Power transmission lines may result in continuous long-term exposures above 10 k V m^{-1}. While epidemiological studies fail to detect a strong correlation in terms of any underlying field mechanism, bioeffects are observed. The fact is that biological effects of E and H fields have not been understood except for their behaviour and function at the macroscopic level. This says more about current experimental inabilities and the lack of theoretical field knowledge at microscopic and smaller domains rather than any actual observations of field interactions whether E, H or A.

E and H fields consist of miniscule field particles called photons. Scientists have since antiquity suspected light consists of tiny particles. In the 17th century Descartes discovered the physical laws governing refraction and reflection, suggesting a wave nature of light. Newton saw light analogous to falling objects in a gravitational field. This set up the dual notions of a wave or a particle, leading to the present unresolved debate within science. Photons were discovered as particles by Einstein in his photoelectric experiments in 1905. During the quantum era from 1927 to the present the photon has been misunderstood, its mass assumed zero, and it has had no known internal structure. Its enigmatic nature during the 20th century is best illustrated via the modern version of the two-slit experiment, where it is seen enigmatically as both a wave and a particle via a diffraction picture that builds up mysteriously photon by photon. There is dispute, too, regarding the biological effects of the E and H fields responsible for many of the advances in the developed world. Modern technology opens ajar a door on seemingly miraculous medical therapies based on biophotons at the epicentre of the cell cycle.

Using SFT the intrinsic mass of the photon has recently for the first time been analytically estimated; the photon is perceived to possess an internal structure. Thus a relatively weak spectroscopy compared with the electron can be associated with the photon. This reveals a mechanism by which atoms and molecules can change their state and shape as the ambient energy density varies in a biological region of interest. This mechanism is seen by examining the ML equations as applied to the simplest atomic structure, the hydrogen atom, where the eigenvalue structure is controlled by the constitutive parameters, ε and μ (permittivity and permeability). Via the photon's role as an atomic binding agent there is a mechanism by which the eigenvalues vary. The quantised motions of the electron, proton and photon inside the atom vary with energy density. To analyse the motions of particles such as electrons and protons possessing an elementary charge **q**, Maxwell's EM equations are written as

$$\nabla \bullet \vec{\mathrm{E}} = \frac{q}{\nu_{\mathbf{q}}} \tag{1.1a}$$

$$\nabla \bullet \vec{\mathrm{H}} = 0 \tag{1.1b}$$

$$\nabla \times \vec{\mathrm{E}} + \mu \frac{\partial \vec{\mathrm{H}}}{dt} = 0 \tag{1.1c}$$

$$\nabla \times \vec{\mathrm{H}} - \varepsilon \frac{\partial \vec{\mathrm{E}}}{\mathrm{d}t} = \frac{\pi}{s_{\mathbf{q}}} q \vec{\mathrm{v}} \tag{1.1d}$$

On their own these four equations form CEM. The Lorentz equation for the forces acting on a particle is

$$\vec{\mathrm{F}} = q\vec{\mathrm{E}} + q\vec{\mathrm{v}} \times \vec{\mathrm{B}} \tag{1.1e}$$

Together with the Maxwell Eqs. (1.1a–d), the ML Eqs. (1.1a–e) form a deterministic system that can be solved analytically in comparison with the Maxwell equations that, as they stand, cannot be fully solved for particular cases. The energy density in a region depends on the volumetric density of the photons that form E and H fields and thus the dynamics of the electron and proton inside the hydrogen atom:

$$\mathrm{d}U = \rho \mathrm{d}V = \frac{1}{2}(\varepsilon \tilde{E} \bullet \tilde{E} + \mu \tilde{H} \bullet \tilde{H})\mathrm{d}V \tag{1.1f}$$

One well-known method of analytically solving differential equations is to substitute functional forms into the equations of motion. In the case of many physical problems, exponential forms may be used as a general solution. Where the motion is known to be periodic, complex exponential forms can be employed as a general solution. One functional form that yields a relativistic general solution for the ML equations is

$$r\,(r_{\mathrm{o}}, \omega_{\mathrm{o}}, r_{\mathrm{c}}, \omega_{\mathrm{c}}) = r_{\mathrm{o}} e^{j\omega_{\mathrm{o}} t} + r_{\mathrm{c}} e^{j\omega_{\mathrm{c}} t} \tag{1.2}$$

where ω_{o} and ω_{c} are orbital and cyclotron angular velocities and r_{o} and r_{c} are orbital and cyclotron radii; the general motion has four unknowns.

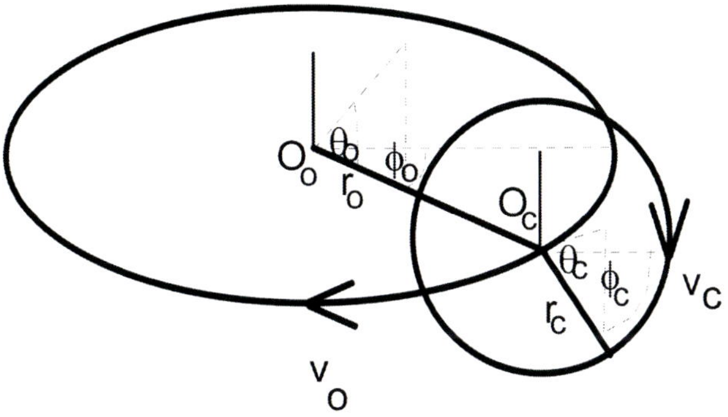

Figure 1.3 The coupled spherical coordinate system at the electron's electric and magnetic centres of motion, coupled via the ML equations. The electron performs a combined motion consisting of two rotations, one orbital and the other cyclotron.

In this form each of the two spinors refers to a centre of motion. For an electron performing a double rotation around an orbital or E field centre of rotation and a cyclotron or H field centre of rotation (Fig. 1.3), its E and H fields can be written as

$$E = \frac{1}{4\pi\varepsilon_0}\frac{q}{r_{\rm o}^2}e^{j\phi_{\rm o}}\hat{r} \tag{1.3}$$

$$H = \frac{1}{4\pi\varepsilon_0\mu_0}\frac{q}{\sin\theta\omega r_{\rm c}^3}e^{j\phi_{\rm c}}\hat{\phi}_{\rm c} \tag{1.4}$$

There is a lot of mathematical history in Eqs. (1.1a–f), and we shall briefly unpick a tiny part of it. Looking at Eqs. (1.1a–d) there are three independent equations in four scalar unknowns. This is in fact the Maxwellian basis of the quantum field equations, which are normally posed as integral equations, a Lagrangian formulation, which is basically symmetric and quadratic. At this stage we cannot solve these equations in only one particle; they are under constrained—too many unknowns and not enough equations. This is the reason for the probabilistic basis of quantum theory. In this form the equations are incomplete. We need to examine pairs of particles that can form a dynamic balance; to solve the equations deterministically we need the added information in Eq. (1.1e). It is this mathematical reality that Einstein tried to point out. Despite the deafening silence of those who had formulated quantum theory he stuck to his view for the rest of his life, with good reason as we will now see.

This important feature of the equations can be understood by looking at the system of deterministic ML Eqs. (1.1a–e) for a pair of particles such as the electron and the proton in the hydrogen atom. If the particles are in dynamic balance Eqs. (1.1a–e) are four equations in four unknowns for each particle, eight equations in all, assuming for the moment that the energy density or equivalently the constitutive parameters ε and μ do not vary in a region of motion. In this form there are four unknowns, and this is supported by the four quantum numbers known to quantum theory. We can use Eq. (1.1f) to convert these equations into six scalar variables, including the E and H field photon densities in any region. The photon densities correspond to the E and H fields and appear in the form of two photon streams, one for the E field and one for the H field, that transit to and fro between the electron to the proton and back again, proton to electron, an overall cycle of photons. Examination of the equations reveals that in any region there are four independent variables and two more dependent variables, including the E and H field streams.

According to the mathematics, if there are any differences across the medium in which the atoms circulate then there will be E and H fields *between* atoms, not just *within* atoms. This can be considered as a dielectric interaction between atoms, a photonic interaction between all atoms; these photons are ubiquitous. The question is ubiquitous: are the micro- and macroscopic fields large enough to cause any observable effects? There are theoretical reasons we investigate in the next chapter to suggest the photon, like the atom, has an internal structure with the ability to change radial or spin states, depending on the ambient energy. Where the energy changes the photon also changes, including sometimes its state, with dramatic effect. Various physical phenomena related to energy changes in a region demonstrate this ability of photons to alter state, thus inducing observable chemical reactions. These phenomena indicate that photons must be present between atoms, a form of molecular aggregate not currently recognised by atomic chemistry. This is part of a 'photon chemistry' overlooked by quantum theory since its inception in 1927. Einstein tried to demonstrate the incompleteness of the statistical nature of quantum theory to no avail. A collection of atoms is balanced by a particular geometric allocation or pattern of fields between atoms, for instance, in a crystal, liquid, snowflake or avalanche; in the formation of lightning, wind, thunder and rain; or in the spin of the sun or the earth. Depending on changes in temperature, the quantum state of the overall system of atoms can change dramatically as the photons change their state. A similar phenomenon occurs at anaphase when chromosomes divide into pairs; a sudden change occurs in the photon state, leading to a molecular reaction as the overall energy state in the cell falls. This occurs as the DNA lines up due to a biogenic E field to act as a coherent liquid crystal. We examine this case in detail when we study cell division in Chapter 5.

A specialised term for the form of solution consisting of two rotating orthogonal vectors, vectors that may rotate in planes at right angles to each other, or planes parallel to each other, is a 'bi-spinor'. The solution is found in mathematics, physics and biophysics, for instance, the shape of the DNA double helix. The bi-spinorial solution is not classical as (1) it treats the photon as a wave *and* a particle, not only a wave as in classical EM nor only a particle as in classical mechanics; and (2) the solution is fractal; there is no limit on size at either end of the size spectrum; an infinite series of solutions running from the universe to the photon is implicit in SFT. We can separate the motions into domains. Physics and biophysics unify under a single theoretical basis where biology and life forms are involved from universal to photonic domains via a family of bosons.

In their equivalent spinorial form the ML equations reveal a mutual system of self-fields inside the atom. The electron that rotates in two orthogonal directions causes two fields that drive the motions of the proton, causing it to also move in these two orthogonal directions; the proton in its turn rotates in the two orthogonal directions, causing two fields that drive the motions of the electron, causing it also to move in two orthogonal directions—in other words a mutual system of self-fields. The two self-fields are streams of photons rather like 'beads on a string' where the photons also have two internal sub-photonic particles that also perform two rotations as they move. This is the essential difference between self-field EM theory and CEM theory; SFT is bi-spinorial behaving relativistically in an intrinsic fashion, while CEM needs further mathematical intervention to demonstrate its relativistic nature.

In Eq. (1.2) only discrete eigenvalues of r_o, r_c, ω_o and ω_c are allowed. After suitable algebraic manipulation, the system may be recast in terms of the energies in diagonal matrix form. The principal mode ($\lambda = 1$) of the resulting eigenvalue system of equations can be written as

$$\begin{bmatrix} 1 & 0 \\ 0 & 1 \end{bmatrix} \begin{bmatrix} V \\ T \end{bmatrix} = \begin{bmatrix} \hbar\omega \\ \hbar\omega \end{bmatrix} \tag{1.5}$$

where $V = V_o = \frac{1}{4\pi\varepsilon_0}\frac{q}{r_o} = V_c = \frac{1}{4\pi\varepsilon_0}\frac{q}{r_c}$, and $T = T_o\ T_c = 1/2m_e v_o^2 = T_c = 1/2m_e v_c^2$ are orbital and cyclotron components of the potential and kinetic energies. In the principal mode $\omega = |\omega_o| = |\omega_c|$ and $r = |r_o| = |r_c|$ where the orbital and cyclotron velocities are also equal $v_o = \omega_o r_o = v_c = \omega_c r_c$. Equation (1.5) contains the elements needed to rewrite the Schrödinger equation for the hydrogen atom in SFT form. Note also that the right-hand side of Eq. (1.5) contains a variable, herein termed Planck's 'number', that empirically agrees with the known value of Planck's constant to an accuracy of seven significant figures. Planck's number $\hbar$ comes from $\hbar\omega = \frac{q^2}{8\pi^2\varepsilon_0}$ so $\hbar = \frac{q^2}{4\pi\varepsilon_0 v_0} = \frac{q^2}{4\pi\varepsilon_0 v_c}$ can be calculated from the solution of Eq. (1.5) that agrees with similar accuracy to the Bohr radius and the known resonant frequency of the hydrogen atom. To see the relationship between Planck's number and velocity, let $\hbar = \frac{E_0}{\nu_0} = \frac{q^2}{4\pi\varepsilon_0 r_0}\frac{1}{\nu_0} = \frac{q^2}{4\pi\varepsilon_0 v_0}$, where ν_0 is the electron frequency and E_0 is the E field potential energy. By involving the effect of the H field upon the electron the solution forms an extension of Bohr's theory. As it stands Eq. (1.5) contains four variables yielding the deterministic motion of the electron given by Eq. (1.2). This agrees with the four quantum numbers known via QM.

Thus for the hydrogen atom the completely deterministic solution agrees with the early Bohr theory, providing a missing magnetic motion, and an analytic expression for Planck's number, the heuristic basis for the eigenvalue equation at the heart of quantum field theory (QFT). SFT can also be applied at the photonic domain to a two-particle model of the photon. The bi-spinorial form again yields a consistent physical model for the photon, including its continuous energy response, its wave–particle duality and the physics of relativity. The time and length dilations of the Lorentz transformation and the self-fields discussed by Abraham and Lorentz are understood in terms of external and internal motions of the photon. Relativity is not applied to space but to the EMF. This is a major difference between Einstein's relativity and SFT. In the cosmological application to universal inflation, there is now a point in space-time corresponding to the big bang unlike the opinion of current physics, which maintains space itself is expanding. In SFT it is only the field, the photon that expands via its radial state; as the energy density diminishes with cosmological inflation the photon expands along with its wavelength. In Chapter 3 we investigate the role of the universe, its gravitational structure and cosmological evolution to constrain biological evolution.

Certain biophotonics mechanisms are apparent:

(1) The relative permittivity and permeability of the tissues and organs of the body vary; the self-fields and the kinetic and potential energies of corresponding particles in the same atom or molecule also vary. These vastly differing dielectric constants induce differences in molecular reactions and energies, most importantly during replication of different tissues and organs. The extracellular material of a tissue, including the number and size of its particular cells, largely determines the reactions and energy states of a tissue's cell cycle. This causes the specific frequencies required for the normal occurrence of the stages of the cell cycles of different tissues to vary from one another, even though similar molecules, chromosomes and proteins are involved.

(2) A phenomenon of cell–cell electric polarisation and alignment is involved in the build-up to cell division. Once the spindle poles are created, charged proteins in the plasma membrane of a replicating cell polarise the external membrane of this cell. This causes the membrane proteins of neighbouring cells to polarise and align, thus creating a feedback to magnify the original E field across the spindle poles within the replicating cell. A complementary process of cell–cell magnetic polarisation is also involved in metaphase and anaphase. This concerns the

magnetic moment of DNA, which at metaphase starts emitting an axial field that influences the rotation of charged particles in the plasma membranes in neighbouring cells, which in turn induces a lowering of the energy within the DNA of the replicating cell. The overall process is a slowly shifting SFT balance between dipolar E and H fields.

(3) As these biogenic E and H fields change, real-time changes to the shape of the DNA molecule follow. Slow changes in energy control the photon states and their binding strength within the chromosomes, leading to structural changes as one distinct photon state leads into another; this is seen during metaphase concluding suddenly in anaphase. This process is similar to the development of storm clouds, the associated lowering of temperature and the eventual production of rain during storms. In the cell cycle, the separation of the chromosome into chromatid halves is achieved via a similar build-up of electric and magnetic energy, lowering the temperature and changing the hydrate shape of the DNA similar to the formation of snowflakes out of clouds.

(4) Biophotons in the form of a north–south-directed E field within a developing cell due to the extracellular dielectric feedback mechanism discussed above flow via microtubules to the DNA. Then the DNA acts as a coherent source of biophotons in the form of an axially directed current that interacts with the membrane proteins in surrounding cells to cause them to rotate, feeding back into the original cell by lowering the ambient energy in the neighbourhood of the DNA.

(5) Resonance plays a major role in the way photons react with biological entities such as DNA, bacteria and viruses. The physics of this resonance is similar to that of a half-wave antenna. A length associated with a molecule, DNA, bacteria or virus may be used to specify a half-wave frequency. In the case of the photon acting as a binding agent inside atoms, the photon is resonant when it supplies a half-wave of phase to its overall transit to and fro within the atom. Such resonances can be used to promote or block processes, including the cell cycles of malignant or healthy cells, or perhaps to act as a bacteriotoxin.

Across the 150 years of EM research and development, it has not been realised till recently that the ML equations would yield exact eigensolutions for the motions of the particles and photons for various configurations of charged particles and regions of varying constitutive parameters. Much of our knowledge for 80 of those years has also been filtered by the probabilistic mathematics of quantum theories. It is only now that the self-fields of systems of dynamic particles such as atoms, molecules and proteins can be determined that the role of the photon is revealed in binding these structures together via a new level of discrete dynamic equilibria. In determining Planck's number is found the reason for the mathematics of the assumed eigenstructure of quantum theory.

For a given dynamics of the sub-atomic particles, there is a corresponding relationship between the E and H fields and the ambient energy density for a many-atom structure. We can see this by examining the variables in Eqs. (1.1a–f). As temperature or some other energy parameter such as pH drops, so too the self-field configuration changes as the photons change state. This can result, for instance, in a relaxation process with the emission of photons. But this may not be a normal random thermodynamic process as given by statistical thermodynamics. The molecular energy states can cluster together rather than being dispersed over a bandwidth of temperatures as with blackbody radiation. In the same way that rubidium crystals act as sources of coherent photons, or lasers, so too DNA and other biological systems can produce coherent photons controlling replication and other biological functions. This coherent behaviour of DNA may well be shape-dependent.

Various researchers are investigating ultraweak photon emissions by biological systems. Strands of DNA, bacteria and viruses have been observed to spontaneously emit photons in the visible part of the EM spectrum, in a range of intensities from 1 to 1,000 photons cm^{-2} sec^{-1}. These observations correspond qualitatively to the theory discussed above; the scientific method requires the measurements also match the theory quantitatively. These findings also correspond to reports of cellular bioeffects due to so-called non-thermal levels of EM radiation exposures, leading to concern among the public and the scientific community. There has been a flurry of research into cellular and other microbiological effects. At the same time research into the medical uses of EMFs has recently entered a new more optimistic era. Researchers in this new area need to examine the role of biogenic photons emitted and absorbed by biological entities and their therapeutic use within the cell cycle. Other areas of research such as genetics, microbiology, evolutionary biology and stem cell technology may also learn from these insights into how cells operate at the biophotonic level.

In many ways biophotonics has had to travel a difficult and complex path over the past 150 years. One of the reasons for this difficult gestation was the minute size and intrinsic energy of the photon compared with other forms of physical interaction, including nuclear energy. Maxwell identified the wave nature of light by expressing his original 20 quaternion equations as 2 wave equations in the E and H fields. He went on to show EM could take the form of a plane wave traveling in space at the speed of light. In the decades that followed long-range radio transmissions validated this wave-like character. In 1905 Einstein identified the photon as a quantum of EM energy via the photoelectric effect. Einstein's experiment had demonstrated a particulate nature of light. Hence there was a question concerning the photon—was it a wave or a particle?

Consequently the photon's physical nature was not understood. In the light of the quantum theory that had emerged by 1927, the photon's fate was sealed for the rest of the century. Heisenberg considered the photon an unknowable entity. Either the photon's momentum or its position could be known, but it was impossible to simultaneously know both; perfect knowledge of the photon was not possible. Heisenberg's uncertainty principle (HUP) was seen as part of the physical fabric of reality. HUP was after all the underlying basis of the commutative equations of quantum theory. Another reason was the experimental fact that the only way to know the photon was doing required another photon, and that would disturb the experiment. One way around such an experimental difficulty is to use our intellects to model the internal motions and structure of the photon exactly as we do in SFT. However, up to the very recent time, the photon was thought a singularity in space.

Another major reason for the difficulties that biophotonics endured during the 20th century can be associated with the general excitement surrounding the era of atomic chemistry that emerged following 1927 and the newly discovered mathematics of the QFTs. Immediately following 1927 a ‘golden age’ of experimental and mathematical research unearthed many profound discoveries of atomic chemistry, culminating with the structure of the DNA double helix macromolecule in 1953 by Watson, Crick and Franklin. In all this work, the role of the photon was not understood. While it was known to ‘mediate’ the EM forces the photon remained an enigmatic wave particle, and its role within the DNA lay dormant underneath the motions of the electrons and protons whose orbits yielded partial information as given by QM and QFTs with their uncertainty.

The level of misunderstanding concerning the photon and radiation generally during this period can be illustrated by the discovery of nuclear energy by Röntgen in 1895. He was testing a vacuum tube in the dark to see if any light could be observed outside the tube. The screen at one end of the tube would emit visible light when struck by a beam of electrons. Unexpectedly a faint glow could be seen on the other side of the room. This was not ultraviolet (UV) light since there was a screen that captured these frequencies. While trying to find out more about this phenomenon his hand somehow passed within the beam and he saw a blurred image of the bones in his hand on the screen (Fig. 1.4). Fearing his colleagues would think him insane Röntgen performed his research in secrecy until he could determine to his own satisfaction what was going on. The phenomenon was immediately realised to be a medical imaging technique that could see inside tissues without surgical intervention.

1.3 Uncertainty: The Incomplete Photon

The Einstein–Podolsky–Rosen (EPR) paradox refers to a paper in which Einstein challenged quantum theory that he saw as incomplete. As a consequence of this historical debate quantum entanglement is now understood as a way in which parts of a QM system connect. The constituent states are linked together so that one part cannot be properly described without mention of other parts within the system. Einstein referred to this as ‘spooky action at

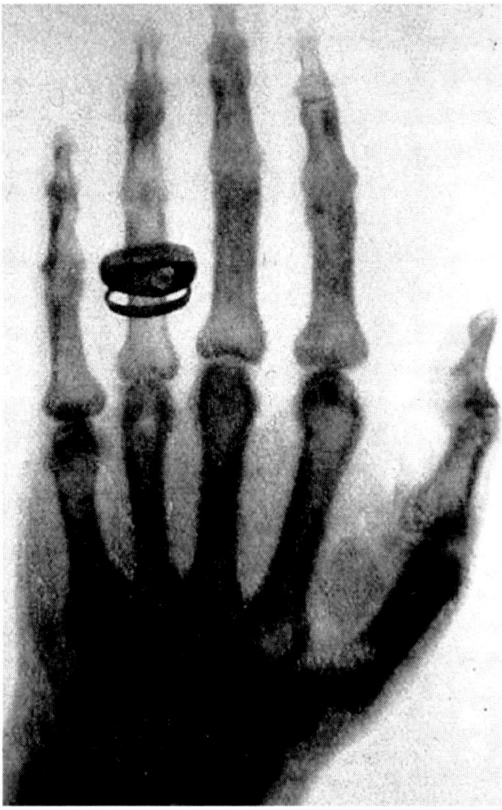

Figure 1.4 A radiograph taken by Röntgen of a hand at a public lecture in 1896.

a distance'. SFT too sees quantum theory as incomplete. SFT suggests a missing coordinate within the photon that is the underlying reason for HUP and its incomplete knowledge at the photonic level. Within atomic and molecular physics according to SFT photons link atoms together. This is the underlying photon level of interaction applying to matter anywhere within the universe where there is an energy difference.

SFT sees two extra quantum numbers that are linked to the discretisation of the paths of rotations of the electron and proton within the atom, which are related to the photon in its role as binding energy. This applies to crystals where atoms rotate in a

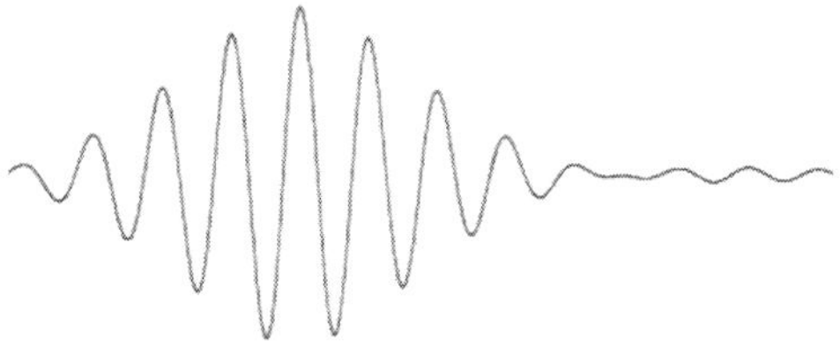

Figure 1.5 Wave packet with uncertainty Δx in position and Δp in momentum.

synchronised fashion and to biological systems where biophotons emitted from DNA acting as a liquid crystal at metaphase can link to other objects in the near field of the biological system. This usually implies the biological capacity for optical sight but can encompass other forms of biophotonic emission and detection. This includes the complete body fields of biological systems in general but is found in specialised physiological tissues such as the elasmobranch fish that use pores, the ampullae of Lorenzini, on the periphery of their dorsal fins for predation. Einstein was right; it is now realised that an interconnection, a binding energy existing within living and non-living systems, has been overlooked; quantum theory failed to see the photonic- and biophotonic-level interactions.

The uncertainty principle was first proposed by Heisenberg in 1927. HUP is now presented in the original form given by Heisenberg. Assume the wave packet illustrated in Fig. 1.5 is a wave packet consisting of sinusoidal plane waves of wavelengths close to λ_0. There are approximately $n = \Delta x/\lambda_0$ wavelengths within the packet. Outside the limits of the wave packet the waves must cancel each other by interference. This only occurs if at least $n + 1$ waves fall inside the limits of the wave. Thus we may write

$$\frac{\Delta x}{\lambda - \Delta\lambda} \geq n + 1 \tag{1.6}$$

In Eq. (1.6) $\Delta\lambda$ is the approximate range of wavelengths. Thus

$$\frac{\Delta x \Delta\lambda}{\lambda_0^2} \geq 1 \tag{1.7}$$

The group velocity of the wave can be written as

$$v_{\mathrm{g}} = \frac{h}{m_{\mathrm{p}}\lambda_0} \tag{1.8}$$

The spreading of the wave is thus characterised by a range of velocities:

$$\Delta v_{\mathrm{g}} = \frac{\hbar}{m_{\mathrm{p}}\lambda_0^2}\Delta\lambda_0 \tag{1.9}$$

By definition $\Delta p = m_{\mathrm{p}}\Delta v_{\mathrm{gx}}$ hence

$$\Delta p_{\mathrm{x}}\Delta x \geq \hbar \tag{1.10}$$

HUP may also be written in the corresponding form

$$\Delta t\Delta E \geq \hbar \tag{1.11}$$

$$m_{\mathrm{e}}v_0^2 = 2\hbar\omega \tag{1.12}$$

The inequality relationship Eq. (1.11) applies to photons, electrons and in fact any quantum system The equation for the orbital motion of the electron lies at the heart of SFT and is given in Eq. (1.12) for comparison. Apart from the approximation of energy and time in the uncertainty relationship and the replacement of the 'greater than' relationship by an exact relationship, the two equations are strikingly alike. SFT completes Bohr theory, which did not include any magnetic effect on the electron.

According to SFT the photon can be modelled as two point masses of opposite charge, similar to the hydrogen atom, but of equal mass. Due to the mass equality the eigenstructure is continuous in this case since a resonance condition holds at every energy state in agreement with Einstein's theory of the photoelectric effect, proposed in 1905, and the equation relating photon energy to its frequency $E = h\nu$ (the Planck–Einstein equation). If such a photon moves past an observation point, the resulting effect is approximately as shown in Fig. 1.6. The wave packet conceptualised around the time of the discovery of quantum theory is thus seen to be an approximation to the physics. The SFT domain of the photon $\omega_\gamma \leq \frac{m_\gamma c^2}{2\hbar}$ or $\alpha/8 \approx 0.912 \times 10^{-3}$ rad sec^{-1} where the inverse fine structure constant $\alpha = \frac{v_{\mathrm{e}}}{c} = \frac{4m_\gamma c^2}{\hbar\omega_\gamma}$ and where the mass of the photon have been estimated[a] $m_\gamma = 0.396 \times 10^{-55}$ kg (0.221×10^{-19} eV). Below this level, possibly in deep space at the edge of

[a] In SFT each photon is assumed to transit between the electron and the proton via a series of resonant elastic collisions. The phase length during transit, $\pi/2$, maintains the atomic periodicity, providing a method for analytically comparing the energy

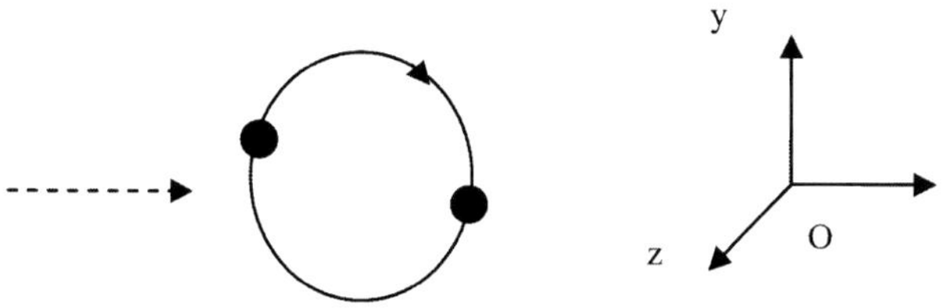

Figure 1.6 Composite photon moving past point O along the x axis rotating in the xy plane.

the universe, the photon's self-energy is significant and it basically 'freezes' and can remain in an isolated state where it performs bi-spinorial rotations about a point in space. Above this energy level the photon's self-energy becomes insignificant and it obeys outside influences to its motion, moving at the speed of light. In this mode the photon acts as a binding energy between inanimate particles and in specialised roles within biological systems.

Looking at Fig 1.7 we might model the wave packet in a somewhat unphysical fashion, as a single particle rotating around a centre of motion while translating along the x axis. In this case uncertainty Δx in the x direction involves a trigonometric function in θ and the scalar radius r of the photon's internal motion. Δx ranges from positive to negative in keeping with the wave motion. Similarly the uncertainty in momentum Δp is also a function of θ and r, as well as the particle's mass, and also ranges from negative to positive. Attempting to investigate uncertainty in this case leads to an unphysical result. It is only when the photon is modelled by two particles, as in Fig. 1.6, the uncertainty function becomes balanced

---

of the photon with that of the electron $m_\gamma c^2 = \frac{\hbar\omega_\gamma v_e}{4c} = \frac{q^2\omega_\gamma}{16\pi\varepsilon_0 c}$, where ω_γ is the integer transit frequency of the photon within each cycle of the electron. For the atom to maintain its periodicity N_γ the number of complete transits per atomic cycle must be an integer. Ignoring any non-linearities, the electron's motion can be modelled as piece-wise linear. This collision-based form of periodic motion by the electron illustrates the non-classical, quantum nature of SFT. Assuming a polygonal motion circumscribes a circle representing the Bohr mageton, the photon collision frequency was estimated as 53 to several places of significance from the known value of the Landé g-factor assumed precisely known from both measurement and quod erat demonstrandum (QED) theory. More refined calculation may be needed to validate N_γ as the orbit is actually non-linear. The experimentally known fine-structure constant can also be used to estimate the solution as $N_\gamma = 54$. This was the solution chosen to estimate the photon mass.

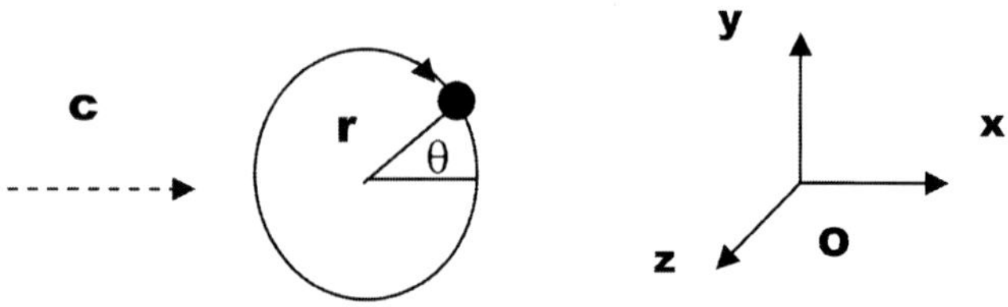

Figure 1.7 A single-particle photon moves past point O along the x axis rotating in the xy plane.

by the symmetry of the sub-photonic particles and uncertainty is obviated. HUP can be considered a theoretical error leading to numerical inaccuracy. SFT is a mutual effect between two particles that obviates uncertainty; the photon motion is deterministic.

The equations of HUP can be seen to be closely related to the central equations of SFT where the motion of the electron is analysed. These equations are exact, whereas in HUP they are inequalities. In this case the motion of the electron is modelled via an infinite-mass proton, similar to the one-particle photon model of Fig. 1.7. In this case the model is not unphysical in the sense that this is a reasonable approximation to the physics using a simplifying assumption. Note that Heisenberg assumes that neither is the group mass a variable depending on relativistic speed nor is it zero. Heisenberg prefaced the derivation of his principle of uncertainty:

> The uncertainty relation specifies the limits within which the particle picture can be applied. Any use of the words 'position' and 'velocity' with an accuracy exceeding that given by equation {(5–6)} is just as meaningless as the use of words whose sense is not defined.... In this connection one should particularly remember that the human language permits the construction of sentences which do not involve any consequences and which therefore have no content at all-in spite of the fact that these sentences produce some kind of picture in our imagination; e.g., the statement that besides our world there exists another world with which any connection is impossible in principle, does not lead to any experimental consequence, but does produce some kind of picture in the mind. Obviously such a statement can neither be proved nor disproved. One should be especially careful using the

words 'reality', 'actually' etc., since these words very often lead to statements of the type just mentioned.

Heisenberg's work revolved around *intuitive methods*; he praised Feynman's graphical method of 1950 as such (*anschauliche methoden*). Quantum theories are essentially numerical models resulting from *imaginal creative thought*, *visualisations* or *visualisability* (see Miller, 1995). It is noted that while uncertainty for any model can be zero this concerns only the model's theoretical basis; this tells us little about how the model stands up to physical experiment. We can conclude that uncertainty is a form of theoretical inaccuracy stemming from faulty intuition as distinct from a numerical inaccuracy or, indeed, a physical inaccuracy. Heisenberg thought the photon could not be understood in a deterministic fashion. While it is correct that any attempt to use photons to examine a photon disturbs any physical experiment, this does not prevent us from mathematically examining the photon. Experimentally we may simply use statistical numbers of photons to test individual photons. When we see blue light in the day sky we are not perturbing *all* photons with blue wavelengths but only those photons that reach our eyes. Hence we assume with near certainty that the vast majority of photons in the sky (near us) will have an associated blue energy. Such insights (intuitions) lead to a formalism of the insides of the photon.

SFT yields exact equations, Eq. (1.12), rather than the inequality equations, Eqs. (1.10–1.11). In retrospect Heisenberg's statements seem more about an afterlife rather than scientific comments about uncertainty being a fabric of reality, which was their context. If one were able to ask one's great-great-grandmother whether she believed in the reality of television, her answer would depend on whether she had ever lived in the age of television. Otherwise she would have denied such a preposterous suggestion, and her view would have been supported by thousands of years of human experience. Similarly, quantum mathematicians might deny the existence of deterministic equations of motion for the electron in the atom. Depending on whether they knew of the modern variant of SFT, they might think physics was a fixed and constant commodity, as immutable as life on earth, or perhaps more pertinent

Planck's 'constant'. Now we can perceive a complete world of theory inside the photon that promises to yield new physical and biophysical understandings. Perhaps this HUP episode of science in the early and late 20th century gives us a more complete understanding of the differences between modelling and reality. We accept the assumptions of a scientific model on faith in order to find a way forwards via theory to an understanding of reality. The more two-way reflections there are between physics and any theory, the more refined and in focus become our perceptions. Ironically we must keep our cerebral antennae open to outside broadcast without closing off reception, as was the case at the 5th Solvay Conference in 1927, when Heisenberg and Bohr suggested, '*We regard QM as a complete theory for which the fundamental physical and mathematical hypotheses are no longer susceptible of modification*.' It is pointed out that such an emphatic faux pas was not the end of the story.

In their normal role biophotons convey energy and information to parts of a biological system, in addition to their role as a binding energy at the EM level. One form of photonic information is phase length, whereby biological systems can infer positions and lengths of their surroundings for various purposes, including sight, touch, smell, motion, location and predation. A body's fields connect it with its environment. These whole-body fields or biophotons appear no different to ordinary connective photons except that they perform an informative role. Within the body biophotons perform this informative role and in addition provide a distribution of energy to its cells, organs and tissues. DNA is central to this role of supplying energy during the cell cycle. SFT suggests self-organisation within a cell occurs because of a reducing level of energy as the cycle proceeds. Biophotons provide information and energy when a group of cells decides to replicate their own species. In time a biophoton may perhaps be found to be as different to a photon as a rolled-up armadillo (Fig. 1.8a) is to a conceptual buckyball of similar size (Fig. 1.8b).

In the light of SFT, uncertainty is seen as an approximation to the physics of the photon, the electron, the nucleus and any system studied using the quantum formulation of the 20th century. The model of the photon as a wave packet is the reason for its numerical

(a)

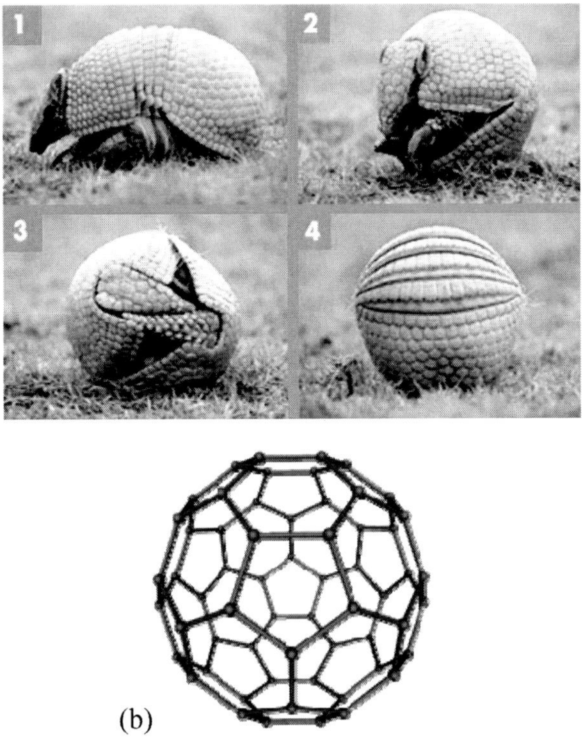

(b)

Figure 1.8 (a) Armadillo rolling up into a ball; (b) buckyball (Buckminsterfullerene C_{60}).

basis. As Einstein long suspected and attempted to demonstrate via EPR, quantum theory is incomplete. It is the exact solutions to the ML equations and the photonic and biophotonic levels revealed by SFT that give quantitative knowledge of physical and biological systems.

As has been demonstrated the mathematics of uncertainty and the equations of SFT are remarkably similar; it remains a historical quirk of fate that SFT did not emerge before World War II (WWII). It may be that the scientific world does not move with any haste. That may be both a good and a bad thing—good in that no presumptuous errors are made but bad in that the world had to wait 80 years to finally begin to peer inside the photon with all the physical and biophysical knowledge that this contains.

One last comment is made regarding the benefits and uses of statistical mathematics and inference. Epidemiology has revealed some of the most medically damaging diseases of the 20th century. The relationship between lung cancer and smoking, asbestos and radon gas was revealed using the methods of epidemiology. The statistical methods used by health specialists today are based on the various censuses conducted many centuries ago, including the Domesday Book in 1086 and the ancient civilisations of Egypt, China, Persia and Rome. When used with knowledge of its limitations and requirements, statistical information of diseases and bioeffects can be the most useful primary tool in the hands of the experienced medical investigator.

1.4 The Role of Science in the Industrialised World

The Industrial Revolution began in the late 18th century in major English cities, including Manchester and Birmingham. The impetus for change towards the modern world was achieved primarily by innovators with cooperation from universities and research institutes covering mechanical and chemical knowledge. Adam Smith's writings on economics along with the prevailing Protestant work ethic gave motivation towards a future ideal of a world of plenty. An increase in life expectancy, a general reduction in working hours and more time for educating children and caring for the elderly are all a direct result of the societal changes within the developed world. The pre-existing servile life of peasants and the majority was such that they were considered in law the chattels of their lords, and if that did not provide personal 'liberty' then the state also had claim on their time, produce and efforts. Thus in the West an almost universal conviction grew that the way forward was via industrialisation.

That factories became sweat houses was an unfortunate historical outcome. Charles Dickens wrote extensively of the child labour, poverty and injustices of early 19th-century England. In 1913 the Ford Motor Company introduced the world's first moving assembly line, which reduced chassis assembly time from $12\frac{1}{2}$ hours to $1\frac{1}{2}$ hours, making cars much cheaper to produce. Such methods of mass production were extremely hard on workers, and Henry Ford would not recognise any trade union, instead using armed police to deal with striking workers. The contemporary social upheavals in Europe that led to World War I (WWI), the October Revolution in 1917, the Great Depression and WWII can all be seen as an ongoing sequence of imbalances of various kinds—social, financial, industrial, national and global. As time passed, the urbanisation that followed industrial progress after WWII also became unbalanced, and the hippies of the 1960s were a counter-revolution seeking the benefits of the pastoral life that had been disconnected from urban dwellers by the Industrial Revolution almost two centuries earlier.

The canals of mid-England are a legacy of the days when water was a form of industrial transport. While coal power drove Watt's steam engine, Tesla and others devised ways of using dams across rivers to provide enormous levels of continuous electrical energy. Dams, including Niagara Falls in 1895, the Hoover Dam in 1936 and the more recent Three Gorges Dam completed in 2006 across the Yangtze River, were all designed to provide hydroelectricity en masse for the industrialisation, urbanisation and globalisation processes that have continued since the start of the Industrial Revolution.

Globalisation has a historically diffuse nature going back beyond the Hellenistic Age, the Roman Empire and the Silk Route in the East. Its modern variant began with the shipping trade between India, China and the West in the 1850s when the British Empire was at its peak. The telegraph, telephone and radio communications played a major role in this process. The telecommunications revolution began in 1901 when Marconi first sent a message across the Atlantic. Since this time the pace of globalisation has accelerated to the point where nations have now become interdependent. A network of trade and finance powers the modern world with its advantages over the pre-existing serfdom. There is a 'cutting edge' between industrial progress and those parts of the globe where many still eke out an impoverished existence as feudal serfs to warlords. Ongoing wars in Iraq, Afghanistan, Pakistan and other regions are areas where modernity and antiquity rub shoulders. The Internet promises to take any advances to a much more pervasive level. Scientific and financial progress now depends on the efforts of researchers from all parts of the globe. Various forms of engineering, including electrical and nuclear engineering, have impacted on modern society, including medicine and biology. The average length of life in the developed world continues to grow. A more balanced global community is now a distinct possibility after millennia of tribal, regional and national battles.

From a mathematical viewpoint it is known that economic systems operate in cycles involving the market. What we are facing at the present time following the global financial crisis of 2008 is a period of stimulus to kick-start the faltering system. It may be that, like an EM balance, financial systems are a balance of two complementary cyclic forces, one is supply and demand and the other is the social benefit and cost to society of a business activity. These two complementary economic 'fields' can be seen as analogues of the E and H fields, respectively; one is a focused field at the consumer level, like an E field, and the other is more diffuse at the community level, like an H field. Thus there are two interrelated economic cycles that combine to achieve growth, recession or sustainability. This use of mathematical physics to model a non-physical situation is similar to the way the quantum theories evolved in the 1920s onwards. Quantum theory was an intuitive, creative method of 'trial and error'. SFT is closer to the actual physics than QFT. Going forward, the global society may find that sustained growth is not possible, not the way we currently define it at any rate. The gross domestic product may perhaps not always be able to be positive once a region is fully developed; it may instead be nearer zero. Development needs tempering in highly developed regions to avoid the risk of ending up like Easter Island, where overpopulation and overuse of natural supplies led to large-scale starvation and a large decrease in population. Looking to the future our present time is similar to the 1870s, when Maxwell discovered the equations of EM. His four equations have a core that can be extended to the nuclear regions of atoms and gravitational fields. The term 'Maxwellian' indicates mathematical correspondences across technologies similar to the term 'Lagrangian' of quantum theory. A major research and development cycle based on the Maxwellian looms in the decades ahead.

1.5 Problems Relating to 20th-Century Inaccuracies

As we have seen science and technology have induced a revolution in man's existential situation. This has resulted in benefits across the gamut of life and society, including personal liberty, medicine, work place standards, leisure, education, entertainment and personal finance. While the list is long there have been corresponding costs, and we need to reflect on where we are going and where we want to go. Many of these costs involve our communal health and quality of life. For nearly a century quantum theory has stood at the pinnacle of our mathematical knowledge of physics. Planck's discovery of quantum physics was revolutionary. The evolution to QM in 1927 solved the immediate problem of analysing the atom. This brought about a *golden age* in atomic chemistry that resulted in the eventual discovery of the double helix of DNA.[a] In the longer term QM came with side issues and problems that could not be circumvented until SFT and its realisation that the photon should be modelled using bi-spinorial mathematics. This brings relativity into the model in a different way than originally formulated by Dirac, who added a term for spin into the Lagrangian. SFT's added parameter of distance within this new form of relativity overcomes the inaccuracies of HUP. There are also problems with relativity itself, including the mistaken concept that it was somehow space itself that warped in agreement with relativity. Instead there is a hidden variable inside the photon. Other than the hydrogen atom QM cannot solve deterministically, and numerical methods are required. Many-electron and many-atom problems are treated using numerical approximations such as Hartree–Fock. Quantum theory routinely involves infinite terms from which a solution is extracted.

Various disciplines of engineering have been limited by the inherent inaccuracies contained in quantum theory in the 20th century, including nuclear, chemical and electrical engineering. Perhaps science of the past century is best exemplified by the production of the atomic bomb to end WWII. Before July 1945 the effects of the weapon were not understood in any detail whatsoever

---

[a] Note the similarity of the piece-wise bi-spinor of the photon in the atom to the piece-wise double helix of the DNA macromolecule within the cell, indicating a fractal physics.

apart from the fact that it was extremely big, bigger than any other bomb in history. Before it was tested it was conjectured that it might even ignite the atmosphere, perhaps blowing away the whole of New Mexico. That Heisenberg and the Nazi scientists who were attempting to match the Allied effort did not develop the weapon was serendipitous. Compared with such energies, 13 ktons of trinitrotoluene (TNT), the photon's self-energy, is negligible, and to say anything else was at that time ludicrous. But this was indeed 'throwing out the baby with the bathwater' in terms of the mathematical physics contained within the structure of the photon. Only by analysing the photon can the nuclear process be calculated, exactly, without numerical error.

Poor air pollution was known before the Industrial Revolution with the use of coal in densely populated cities such as London but became more pronounced in the late 18th century. That urban air quality is very poor in developing countries such as India and China is testament to where these nations sit on the development ladder. Where large populations are being raised out of poverty air quality is unfortunately not seen as a high priority among present aims. Many domestic, industrial and agricultural chemicals were developed in the 20th century in the era of quantum theory with its uncertainty. The developed and undeveloped worlds continue to suffer chemical pollution of various kinds as tragic cases of dichloro-diphenyl-trichloroethane (DDT) and thalidomide attest. Pharmaceutical side effects are a fact of life at present. Rising levels of male infertility have been linked to water pollution via chemicals that inhibit testosterone, including cancer treatments, and pesticides used in agriculture make their way into the water system, inhibiting fertility across species. Other chemicals mimic oestrogens, causing male fish to change sex. The techniques of recombinant DNA, sometimes called genetic engineering, are seen by proponents as ways to treat diseased humans (gene therapy) and improve food. It is seen as a possible form of genetic pollution by opponents, with many countries preferring to have a moratorium on genetically modified food such as maize wheat and canola to await further evidence of safety. Stem cell technology is much lauded as a therapeutic methodology for a number of diseases, including leukaemia, with a number of other diseases, including cancers, multiple sclerosis and motor neuron disease (amyotrophic lateral sclerosis [ALS] in the U.S.), being actively studied. It has unfortunately raised ethical concerns over using embryonic blastocysts, which opponents, including major religions, believe is a viable human entity. Perhaps induced pluripotent stem cells may be a satisfactory alternative. In all these mainly chemical areas of research the roles of E and H fields are completely unknown. Specialist and undergraduate texts on molecular biology, for instance, lack terms in their index on E or H fields and lack descriptions of how the biophoton interacts below the current chemical mechanisms in each of these cases.

Unlike other forms of pollution EM (sometimes termed 'electromagnetic radiation' [EMR][a]) is invisible; in addition the internal tissues of the body do not in general feel pain as opposed to the skin. The impact of modern technologies on human health, such as the mobile phone and transmission towers, is in question and depends on how the EM exposure impacts the body. At present the standard definition of exposure and dose appears of limited practical use. The current standard of allowed EMR is expressed in terms of a unit called the specific absorption rate (SAR) that is measured over a macroscopic cube of tissue (1 cm^3). Overall effects producing a rise in temperature of 1°C are currently considered hazardous. The trouble with this measure is two-fold. First, only thermal effects are recognised. As Planck's blackbody shows all frequencies are thermal, the premise is correct—only thermal effects do indeed exist, but the frequencies and effects that do not produce a macroscopic measure of heat cannot arbitrarily be ignored as ineffectual such as those at extremely low frequencies (ELFs) or those that produce effects other than a rise in temperature. Second, cells and other biological systems are <10 μm in size (volume $\approx 10^{-9}$ cm^3), so they cannot produce a 1°C increase over a 1 cm^3 cube of tissue. Cellular effects thus fall through the cracks of the EMR standards. The main issue with EMR standards methodology at present is its classical EM basis. So-called 'non-thermal'[b] effects

[a]The term 'electromagnetic radiation' (EMR) is used for health hazards, distinct from the term 'electromagnetic field' (EMF), which applies to health benefits implying a lack or a presence of control over far fields.

[b]We use the term 'non-thermal' as it remains a recognised term within the bioeffects community. However, no demarcation exists apart from microdosimetric limitations. Not until methods are devised that can at the very least measure photon-level fields

are not understood since no other physical mechanisms apart from thermal effects have been recognised. Without understanding the photonic-level mechanisms involved in biology present-day bioeffect studies cannot hope to reveal problems. SFT may well impact both the theory and the experiment of EMR bioeffect studies by providing plausible biophotonic mechanisms of cellular biology.

The list of issues goes on. Inner-city living is another cause of pollution that afflicts all corners of the developed world, with rising rates of crime, stress, noise levels and other forms of urban problems. While urban renewal projects are able to minimise some of these effects, outlying commuters need to travel to city centres. Traffic and noise are endemic problems in densely populated cities such as Mexico City and Los Angeles. Commuters spend a sizeable proportion of their lives in congested auto traffic. Acoustic standards are needed in the same way that EMR standards have developed since the end of WWII. Other forms of pollution such as marine pollution exist. The use of sonar technologies in the marine environment needs standardisation. The bleaching of coral reefs threatens areas of biodiversity, such as the Great Barrier Reef. The most tragic instance of misuse of scientific technology is the drying of the Aral Sea due to Soviet era irrigation (Fig. 1.9).

The late 20th to early 21st centuries may well be seen as a turning point in human history as time goes on. That man came to edge of the abyss and looked over is the reality of the atomic age that followed WWII. Mutually assured destruction (MAD) will stand out in human history as the peak of man's insanity as the world went to war for a second time, barely catching breath before entering the Cold War. Far from being the godsend that in reality it is, nuclear power has been a trial for humankind thus far in its development. The 20th century may have been an 'approval' period for mankind in our quest for safe, clean nuclear energy. The mathematical solutions and inherently more accurate physics within SFT may help revolutionise EM and nuclear physics, allowing closed-form solutions to applications, thus refining the engineering. A similar impact may be felt within biology, medicine, pharmacology and

---

inside cells will measurements catch up with current biophotonics theory or SFT cellular theory.

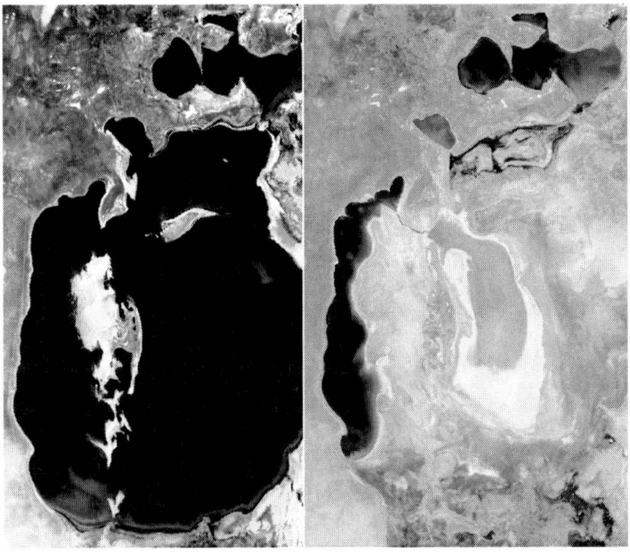

Figure 1.9 Drying of the Aral Sea due to Soviet irrigation diversion (credit: Wikipedia).

microscopy (where the living cell[a] is 'stained'—effectively stabilised but perturbed from its living state).

1.6 Scientific Limitations and Presumptions Pre-SFT

Having been introduced to the basics of SFT we now examine three pre-SFT scientific presumptions and limitations that impact medical knowledge. In each case they relate to the lack of any field-level interaction within the specific view of biophysics and chemistry. The

[a]There is current debate regarding the efficacy of in vitro studies as compared with in vivo studies that is in some ways similar to the current problems of imaging moving, living entities. These problems can be better understood via a photonic-level understanding of E and H fields and how they operate as binding energies within and among atomic and molecular structures. In vivo studies denaturise or perturb the supposedly living entity, either removing it from its environment or devitalising its biological environment.

first of these is the long-standing negative attitude within Western science and medicine to homoeopathy, the second is the attitude of some influential scientists within the bioelectromagnetics (BEM) community to so called non-thermal effects and the third is the complete lack of knowledge of the intricate and intimate role of E or H fields in the many processes of developmental biology, evolution and genetics.

1.6.1 *Homoeopathy*

Homoeopathy is an alternative medicine proposed by Hahnemann in 1796. The fact is that it is used fairly widely in the Western world. It seems that in general it 'works' according to its adherents, including the patients who use the heavily diluted preparations. One particularly influential community of users is the British royal family. Sceptics and some scientists have arrayed their collective forces against this alternative medicine. The basis of homoeopathy is presumed to be the 'law of similars'. This appears like a form of immunology, yet the actual basis for homoeopathy at this point in time is not clear. From Wikipedia:

In 1987, French immunologist Jacques Benveniste submitted a paper to the journal Nature while working at INSERM. The paper purported to have discovered that basophils, a type of white blood cell, released histamine when exposed to a homeopathic dilution of anti-immunoglobulin E antibody. The journal editors, sceptical of the results, requested that the study be replicated in a separate laboratory. Upon replication in four separate laboratories the study was published. Still sceptical of the findings, Nature assembled an independent investigative team to determine the accuracy of the research, consisting of Nature editor and physicist Sir John Maddox, American scientific fraud investigator and chemist Walter Stewart, and sceptic and magician James Randi. After investigating the findings and methodology of the experiment, the team found that the experiments were 'statistically ill-controlled', 'interpretation has been clouded by the exclusion of measurements in conflict with the claim', and concluded, 'We believe that experimental data have been uncritically assessed and their imperfections inadequately reported.' James Randi stated that he doubted that there had been any conscious fraud, but that the researchers had allowed 'wishful thinking' to influence their interpretation of the data.

It is a sad epilogue that Benveniste died in 2004, unable to convince science of his results that gave credence to homoeopathy pointing towards a mechanism. Nevertheless homoeopathy continues to be used on a widespread basis, even with little support from science. Given Benveniste was an immunologist SFT suggests homoeopathy to be a field interaction able to cause immunological response or antibody–antigen chemistry to take place, as suggested by Benveniste's experimental efforts.

Atomic chemistry is governed by Avogadro's number (6.02×10^{23}) of constituent atoms in 1 mole. Hence it is said that the original constituents are lost if there are 12 or more dilutions. Each dilution ends with a 99% reduction, where 1% of the original constituents remain. However, such calculations based on atomic chemistry ignore the permutations of the photon chemistry within a dilution. There are many, many more ways remaining after 12 dilutions whereby the photons are present within the solution in some distribution (hence the shaking). Within water there are an almost infinite number of fundamental geometric orientations (or photon states) connecting molecules. A similar form of photon chemistry occurs within the *spine of hydration* and within the various structures of DNA in the form of chromosome, chromatid or chromatin. When flying with various Chinese airlines there may be a video played before landing of calisthenics to be performed by passengers in their seats that are designed to release tiredness and/or soreness induced by long journeys over several hours or more sitting in the one fairly cramped seat. What may be happening is that such constrained posture 'traps' the body's tissues and cells within certain angled structures. The muscles, bones, cells, etc., can be elastic or rigid, forming a structural system known collectively by the term 'tensegrity'. Energy in the form of photon chemistry may be trapped within the structures of hydration around such anatomical and physiological entities. These can be released by stretching exercises or by other processes involving the system's energy. The effects of old age, rheumatism, aches and pains can be relieved by similar types of stretching exercises; the energy is released via fibres within the body running from the internal cells,

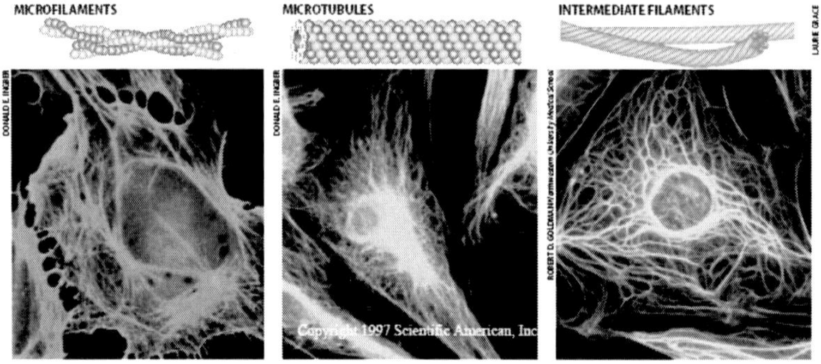

Figure 1.10 Cytoskeleton of a cell consists of microfilaments (left), microtubules (centre) and intermediate filaments (right), all of which are nanometres wide. The rounded shape near the centre in each of these photographs is the cell nucleus. The three components interconnect to create the cytoskeletal lattice, which stretches from the cell surface to the nucleus (top left). The molecular structure of each component is shown above the corresponding photograph. (*Source*: Ingber D.E., *The Architecture of Life*, Scientific American, January 1998.)

tissues and organs of the body, including the motor cortex, to the exterior of the body (Fig. 1.10).

1.6.2 *Thermal and 'Non-Thermal' Effects*

Standards for E and H fields of both public and occupational exposures are controlled by national and international standards committees. Radiofrequency (RF) exposures are linked to the macroscopic E and H fields and the consequent absorption and distribution of RF energy within the body that is strongly dependent on body size, shape, physiology and anatomy and on the incident radiation frequency and polarisation. To limit thermal effects in tissue, SAR is measured in watts per kilogram of tissue. A specific rise in body core temperature of about 2.2°C is taken as the limit of endurance for clinical trials. With respect to RF radiation, a limit of increase of 1°C in rectal temperature is taken as a basis for determining a SAR limit for human exposure. Most standards are based on a SAR of 4 W/kg divided by 10 to give a further safety margin. Thus the general basis is 0.4 W/kg. Unfortunately, the SAR is

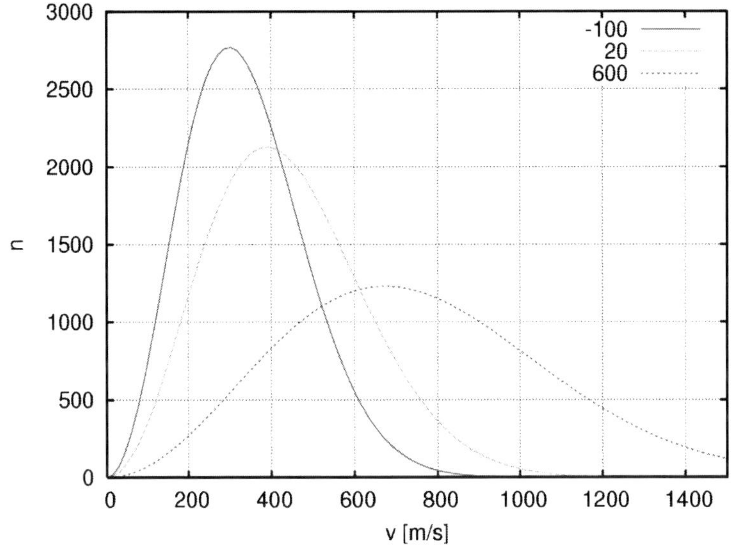

Figure 1.11 Maxwell–Boltzmann distribution for thermodynamics.

not easily measured and cannot be used for practical measurement work. Therefore, derived measurable safety limits based on the SAR are determined in terms of macroscopically measurable quantities such as power density, E field and H field strengths and induced and contact currents that are averaged over a tissue size of 1 cm^3.

When Planck first discovered the connection between thermal energy and frequency of a blackbody, he solved a two-pronged mathematical problem. He rationalised wavelengths across the spectrum, choosing discrete frequencies rather than a continuous range of frequencies. The pre-existing mathematics of the time had been able to separate effects into short and long wavelengths. Given the contemporaneous discovery of radioactivity there was a considerable state of flux in science. The photon's discrete behaviour emerged with the failure of classical science to provide a consistent theory for the energy of a blackbody cavity. Wien at short wavelengths and Rayleigh and Jeans at long wavelengths had experimentally obtained differing analytic equations for the energy. The Rayleigh–Jeans equation indicated an 'ultraviolet catastrophe', an infinite energy at wavelengths around a micron. Planck modelled the blackbody walls, its atoms, as EM dipoles and used the potential theory formulated by Hertz, along with the concepts of probabilistic thermodynamics developed by Maxwell and Boltzmann (Fig. 1.11). Planck had first acted in an 'act of desperation'. He slowly came to realise the theoretical implications of the effect that was an observation of a new quantum physics that Einstein supported via his photoelectric experiments. Fourier's discrete mathematics was now supported by blackbody radiation. The Rayleigh–Jeans law for low-frequency radiation intensity emitted by a blackbody is

$$I_{\mathrm{RJ}}(\nu, T) = \frac{8\pi\nu^2}{c^3}\mathrm{k}T \tag{1.13}$$

The Wien approximation at short wavelengths is

$$I_{\mathrm{W}}(\nu, T) = \frac{2h\nu^3}{c^2}e^{-h\nu/\mathrm{k}T} \tag{1.14}$$

Planck's quantum law is

$$I(\nu, T) = \frac{8\pi\nu^2}{c^3}\frac{h\nu}{e^{\frac{h\nu}{\mathrm{k}T}} - 1} \tag{1.15}$$

For over a century Planck's law has been used to describe thermal interactions with EM frequency. Yet some within the BEM community still claim there are no low-frequency mechanisms or bioeffects of medical concern. One impediment has been that quantum theory has not seen any biophotonic interactions. On the contrary SFT strongly indicates there is a photonic level of interaction across physics, while measurements other than those mandated via exposure standards indicate effects at so-called 'non-thermal' frequencies. Yet Planck's law shows that *all* frequencies are thermal. The assumption within some quarters that non-thermal effects cannot occur below some thermal threshold is based on macroscopic measurement methods and the use of CEM theory that sees no photons within atoms and molecules. At the photonic level there is no thermal threshold. In a recent analysis Weaver et al. examined epidemiological reports of weak field effects via a temperature-dependent function $J\ (T) \approx n\,(T)\,e^{-U_0/\mathrm{k}T}$ with units of molecules per unit time. In Eq. (1.1f) the energy of E and H fields is clearly associated with the volume densities of photons. Weaver's thermal analysis does not account for any photon-level effects. It is no surprise such heat models fail to validate any field effects. This lack of interaction obviates any photon mechanism. Planck's law shows how to include photons within heat analysis, yet in some heat analyses only atomic-level currents are examined. It may be thought changes in molecular structure due to photon state do not induce macroscopic effects such as chemical cascades. Like homoeopathy this may be 'throwing the baby out with the bathwater'. In 1999 Adair wrote for the US Commission for Scientific Medicine and Mental Health:

> **The Physics of Alternative Medicine: The Fear of Weak EM Fields**
>
> One need not be a physician to conclude that the collision of **falling leaves with human heads cannot constitute a significant cause of fractured skulls**. Only to someone who knew only that leaves are parts of trees might it seem even plausible that leaves break heads. But few of us understand magnetic fields as we do tree parts. Are the minute magnetic fields from our power distribution systems that some have associated with cancer leaves or tree-limbs? I answer, 'They are metaphorically leaves, and it is no more possible that they cause cancer than that real leaves crack skulls.'
>
> **Field Basics**
>
> Electric and magnetic fields act on matter through forces on electric charges. The fields can be significant biologically only if they change the energies of charged biological elements as much as the mean energy from thermal agitation, kT, where k is Boltzmann's constant and T is the absolute temperature (about 310 degrees Kelvin). The changing of magnetic fields happens primarily through the electric fields associated with that change (the Faraday effect). Electric fields alleged to be carcinogenic and generated in humans by the 60 Hz 5 milligauss (mG) magnetic fields from an electric power distribution system will be only about ten millionths of a volt per meter (V/m) and cannot induce an energy transfer to biologically significant molecules greater than one-millionth kT.
>
> While there may be **biological amplification mechanisms** that we do not now understand, such mechanisms can work only if the 'signal' is larger than the electrical noise-and the body is electrically noisy. The fundamental random (Johnson–Nyquist)

electric fields from the thermal agitation of charge, acting on cells, are thousands of times greater than the electric fields produced by the 5 mG fields of our environment. Hence the environmental fields cannot affect biology on the cell level.

Over larger regions, that thermodynamic noise is less important as the thermal fluctuations of the charge average out, but there are other endogenous sources of electrical noise. The familiar electric fields generated by heart action, measured in the course of an electrocardiogram, are more than a hundred times greater than the environmental fields. Even over the frequency band of about 60 Hz, the electric fields from the heart have been measured (by John Bergeron) to be about ten times greater than the electric fields from 5 mG 60 Hz magnetic fields.

A young woman appeared on television early in 1993 to express her conviction that the cancer found in her child was induced by power-line magnetic fields while the child was in utero. But the baby was subject to between ten and one hundred times stronger fields by her mother's heart action while her mother was carrying her.

Of course, large electric fields do have biological consequences. Five hundred deaths a year in the United States are attributed to accidental electrocutions. About 100 V/m may cause lethal cardiac fibrillation. Similar currents can restart heart action after cardiac arrest, and smaller electric fields regulate heart action and excite other muscular action. The weakest fields that are definitely known to generate biological effects in humans are the fields of about 0.2 V/m that act on the dark adapted eye to generate visual phenomena (phosphenes). There have been claims for other effects—such as bone healing—at fields as small as 0.1 V/m that are plausible, but not incontrovertibly established. Hence, the fields of 10 millionths V/m, from the electrical power distribution system, are about 10,000 times smaller than the smallest fields known to (harmlessly) effect humans.

Direct magnetic effects are also possible. Bees, some fish, and perhaps birds and other animals navigate by use of compasses of magnetite (lodestone) crystals imbedded in their cells. But at 60 reversals a second, the magnetic forces cancel out and the energies transmitted to magnetic elements in animals by 60 Hz, 5 milligauss, fields can be expected to be less than 1/10,000 kT. Neither birds, bees, fishes, nor humans can even detect such weak 60 Hz fields, let alone be harmed by them.

However, there are many laboratory experiments (I have heard that there are up to one hundred) that purport to have demonstrated effects of very weak fields on cells in vitro. All have been accepted without a rational, accepted model for biological interactions of EMF. Are they all wrong? If so, why?

The probable answer is that experimental errors have been accepted as real effects. Error explains the incoherence and lack of division of the positive reports. It also explains the almost universal lack of a dose-response relation. After more than 20 years of such studies, no well-defined, replicated demonstration of the effects of very weak fields has emerged. Error is the best explanation for all observed reports. (Two workers described in the early 1990s a threefold increase in Myc oncogene expressed RNA on exposure to a low-frequency electromagnetic field, which was then implicated in carcinogenesis. The results could not be replicated by two other groups, which had tightened controls and certain calibrations. The repeat work was reported in the Journal of Radiation Research, October 1995, and reviewed in Science, September 1995. The original erroneous report was widely disseminated in the press and reawakened interest in an EMF/cancer causal relationship.)

But if there is so much smoke, is there not fire? Too much smoke with no sight of flame, suggests to experienced scientists that the smoke is only fog. Seven years ago, I served as chairman of a committee that met in Salt Lake City to report on the National Cold Fusion Institute. At that time, there were 100 papers, from 10 different countries, reporting results that were interpreted as evidence of cold fusion. But cold fusion has been tossed into the dustbin of discredited science; there is no cold fusion.

Flawed Studies

Epidemiological studies that claim to demonstrate effects of weak electromagnetic fields have had great public impact because the techniques and results can be stated in (deceptively) simple ways. In fact, much of the work reporting positive effects is critically marred by errors in technique or analysis and none is even nearly definitive.

As an example of the deficits of a relatively good study, I consider the heralded Swedish measurements of Ahlbom and Feychting that have been cited as showing that the magnetic fields from power distribution systems in Sweden considerably

> increase the probability of childhood leukemia. Among the many malignancies of adults and children living near power lines that they recorded, they reported significant elevations only of leukemia among children who lived in houses that were calculated to have received, on average, more than 1 mG from main power lines over a long period of time; they found 11 cases where they expected 4.5 from their control group. But they also measured the fields from all sources in the houses-though for a brief time-and found only 5 cases where the measured fields were greater than 1 mG. They expected 12.5 from their controls. These are not convincing data. Advocates of the EMF/cancer relationship advertised that 'calculated' fields cause leukemia. The equally significant result that 'measured' fields prevent leukemia was not advertised. Of course, the minute electromagnetic fields—either calculated or measured—neither cause nor prevent cancer.
>
> The Swedish result, accepted as advertised, requires that magnetic fields from the use of electricity are responsible for most childhood leukemia. But the leukemia rate of 4 per 100,000 childhood years is the same everywhere in the West—in the United States and Europe—independent of variations in the use of electricity. Moreover, in the U.S., the incidence has not changed significantly for at least 50 years, while electrical power usage has increased more than 20 times during that period.
>
> Indeed, a National Research Council, National Academy of Sciences, review that considered all epidemiological evidence concluded that 'magnetic fields measured in the home . . . have not been found to be associated with an excess incidence of childhood leukemia or other cancers.'
>
> In summary, there are very good reasons to believe that weak electromagnetic fields from our electrical distribution system have no biological effect at all. And there is no good reason to believe otherwise. The fear of weak magnetic fields cost the U.S. an estimated $23 billion by 1993 and continues to cost in unnecessary transmission line relocation, abandonment of structures, and loss in property values, etc. The EMF/cancer non-problem should be scientifically relegated into the abyss that has swallowed cold fusion, N-rays, polywater, Lysenko's anti-genetics, and other aberrations.

Having read Adair's comments one question springs to mind: If we think of EMR as metaphoric leaves breaking metaphoric heads, then why are we all asked to turn off our mobiles when flying in planes? A mobile phone is miniscule compared to a Boeing 747. The reason given by the pilot is that the EMFs can interfere with the navigation systems of the airplane. Ironically this potential misalignment is similar to what can occur when cells are exposed to sufficient levels of EMR. EMR can interfere with the naturally high-Q systems (Adair's *biological amplification mechanisms*), cells that have evolved to induce mitosis. If electric and magnetic systems are perturbed,, for example, unnatural rotation of components, mitosis cannot proceed. Further, depending on the stage of embryonic development or tissue and organ mitosis that is taking place, *there is no totally safe level* on the basis that perturbations to growth and development should be avoided. This is the same conclusion found for nuclear radiation. At nuclear energies it is the risk of DNA damage; for EMR the DNA is not broken but it can be perturb mitosis outcomes from their unexposed state and this, like nuclear energies, may be due to a single quantum of energy. This applies both to healthy mitotic processes and unhealthy ones. So it can go both ways. But living in environmental levels of EMR and nuclear radiation, a practical basis of technological safety is to make levels as low as reasonably achievable (ALARA).

The 'leaves on heads' brings up another analogue to EMR. In WWII both the Allies and the Axis forces had many occasions to blow up bridges to disrupt traffic carrying supplies, etc., across the bridges. We need not destroy every vestige of the bridge just to stop the traffic from crossing one side to the other. Looking at this as a problem in energetics, we can begin by taking a look at the length of a particular bridge. If we had access to unlimited energy, for instance, we were able to set the wind shear to a predetermined frequency and amplitude, then we could calculate the oscillatory vibrational mode. In the case of the bridge shown in Fig. 1.12 we could choose the longest span, the one on the left, as this would be the lowest energy needed. However, any vibrational shear force would still need to be enormous and at the precise resonance frequency, probably well beyond natural winds. But this is not the only way to stop the traffic on the bridge. We might instead look to bring down the vertical struts on the main bridge. Now we only need a small amount of TNT to get the job done. Destroy one or two

Figure 1.12 Bridge (credit: http://www3.telus.net/pellegrin/Bridge/Bridge.html).

of these critical struts, and no ordinary driver would dare venture to cross the bridge. In other words we only need to attack the bridge at its weakest point. Again if we wanted to destroy a building, we could if it had keystone arches in its architecture; simply remove the keystones, and the arches would tumble down, for instance, the Coliseum in Rome. The key then to disrupting a cell is the architecture involved in a cellular process.

A pertinent example of the above discussion to cancer treatment is as follows. According to current estimates there are around 500,000 melanoma cases in the U.S., causing roughly 8,000 deaths per year. It is claimed that to disrupt DNA the use of relatively high-energy THz frequencies or higher are needed. Note that THz frequencies are around the energy of the UV frequencies observed to be emitted by strands of DNA on relaxation, so such energies correspond with the biogenic mechanism. Recently an RF treatment for cancer using dipoles operating at 100–300 kHz was described by Kirson et al. They performed in vitro and in vivo trials, taking time-lapse microphotography and performing finite-element simulations to record and investigate their findings that they could disrupt the mitosis of melanoma cells. Three main effects occurred:

(1) In treated cells, mitosis began normally but took variable periods before cleavage into two daughter cells.

(2) About a quarter of the cells undergoing mitosis were destroyed as the formation of the cleavage furrow approached complete cell separation. During this process, the cell membrane ruptured, and many small membrane blebs formed, resembling post-mitotic apoptotic cell death.

(3) In treated cultures nuclear rotation was observed. In early mitosis, after cell rounding, the nuclei were observed rotating within the cell. A full rotation lasted on average 15 minutes. This effect resembles whole-cell rotation observed during exposure to intermediate-frequency alternating E fields and reported in the early 1980s (Figs. 1.13–1.14).

Biogenic mitosis should normally occur in the absence of exogenous fields. This process is an SFT balance at metaphase that finally becomes unbalanced leading to anaphase. Cell rotation and axial H fields emitted by DNA are involved in these SFT balances. There is a homeostasis that is maintained during mitosis followed by a gentle build-up of electrostatic and magnetostatic fields, allowing microtubule and microfilament structures to grow until the chromosome cleaves into its two identical chromatids. If perturbing structural rotations occur, exogenous E, H, or EM in nature, the whole mitotic process may be undermined, delayed or prevented entirely. Homeostasis aids the development of cytoplasmic structures that can be perturbed by exogenous fields. Although the melanomas cells discussed above are found within the epidermis, similar effects may occur deeper within the body in cells that are developing. The depth of interaction depends on the constitutive parameters of the various tissues and microscopic structures within and surrounding the cell. One situation of critical importance is the fields found within the developing foetus, as illustrated in Fig. 1.15. Another is the development of breast tumours in teenage and adult females in the tissues of the breast, also shown in Fig. 1.15. Other situations of organs or tissues at risk near the surface of the body are the formation of testicular cancers and brain tumours due to use of radar or communications devices held close to the respective body part. Critically, do exposures reach the developing foetal, breast, testicular or cranial tissues and cells? What level of exogenous fields is able to cause significant effects

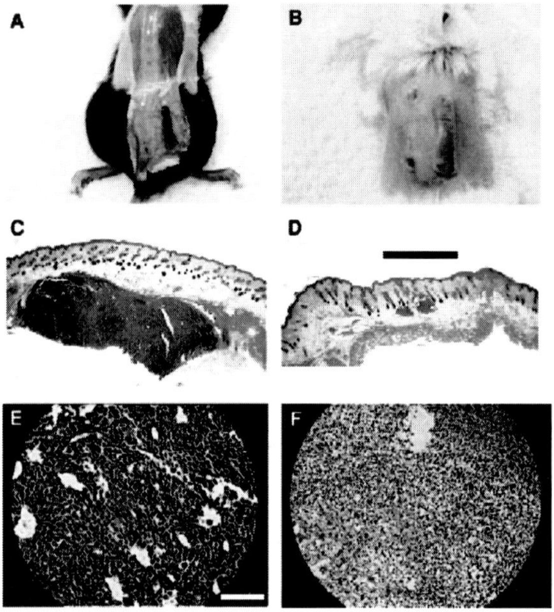

Figure 1.13 In vivo effects of TTFields on intradermal tumours in mice. Malignant melanoma (A) and adenocarcinoma (B) tumour cells were injected in two parallel locations intradermally on the back of each mouse. Only the tumour on the left side of the mouse was treated. After four days of TTFields treatment (at 100 kHz), no tumour can be discerned on the treated side, whereas on the untreated side a large tumour has grown. (C–F) Histological sections of TTFields-treated intradermal melanoma versus a control (untreated) melanoma on the same mouse. (C) After H&E staining, a large (5 mm diameter) nodule of melanoma cells can be seen in the dermis of the control tumour (×40). Note that due to the large size of the tumour, its deep portion has been lost in preparation. (D) Treated tumour; only two small (<0.4 mm diameter) nodules are present (scale bar: 0.5 mm). The non-tumour structures of the dermis are morphologically intact. (E) Control tumour; malignant melanoma cells appear intact and viable (×200) (scale bar: 100 μm). (F) Only necrotic tissue and cellular debris are seen in the treated tumour. (Credit: E. D. Kirson,[1] Z. Gurvich,[2] R. Schneiderman,[2] E. Dekel,[3] A. Itzhaki,[4] Y. Wasserman,[1,4] R. Schatzberger,[2] and Y. Palti.[2] Disruption of cancer cell division by alternating electric fields. [1]Dept of Biomed Engineering, NovoCure Ltd., Haifa, Israel; [2]B. Rappaport Faculty of Medicine, Technion–Israel Institute of Technology, Haifa, Israel; [3]Dept of Molecular Cell Biology, Weizmann Institute of Science, Rehovot, Israel; and [4]Elisha Medical Centre, Haifa, Israel; *Cancer Research*, **64**, 3288–3295, May 1, 2004.) *Abbreviation*: H&E, haematoxylin and eosin.

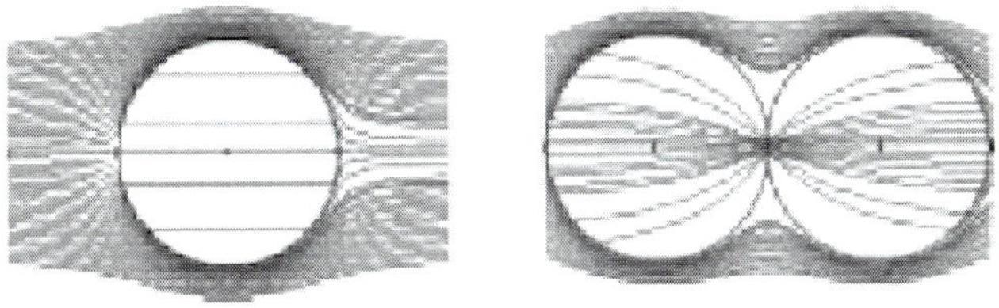

Figure 1.14 Numerical model of currents and fields inside a replicating cell.

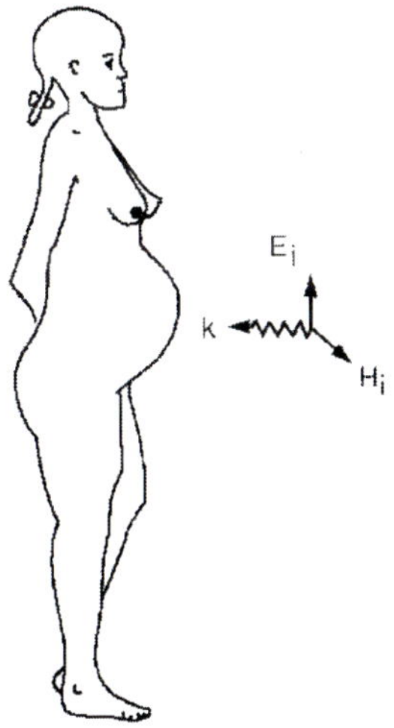

Figure 1.15 Foetal and breast exposures.

upon the otherwise unexposed developmental homeostasis and the normal growth of microstructures? If microtubules cannot develop normally mitosis cannot occur in the normal biogenic time frame. This question needs addressing via in vivo trials and numerical simulations. At present the various national and international EMR standards and many in the BEM community recognise only so-called 'thermal' effects. Another issue that needs study is the use of acoustic or vibrational energies rather than EM energies; the use of sonar appears more able to reach deep-tissue tumours.

Shown in Fig. 1.16 are fractal structures associated with the various forms of DNA. The size of the chromosomes varies across and within species. Chromosomes vary in length from around

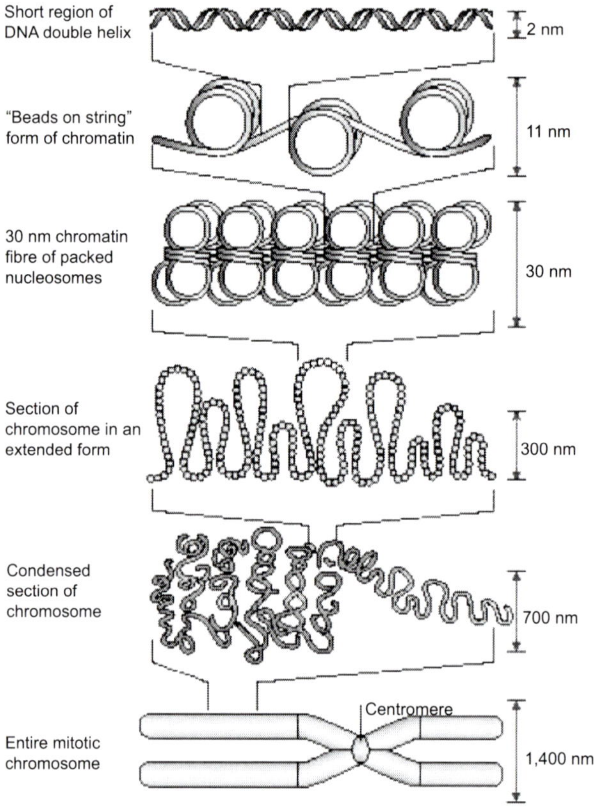

Figure 1.16 Various fractal structures of DNA (credit: Belyaev I., Non-thermal biological effects of microwaves: current knowledge, further perspective and urgent needs, Stockholm University, Stockholm, Sweden; General Physics Institute, Russian Academy of Science, Moscow, Russia, from the workshop *Do Sinusoidal versus Non-Sinusoidal Waveforms Make a Difference?* Zurich, Switzerland, 17–18, 2005).

0.1 μm to 30 μm and across arms of the chromatids from 0.2 μm to 2.0 μm. EM frequencies in free space associated with these lengths range from 150 THz to 1.5 PHz. Cells range in size from around 10 μm to 100 μm. EM frequencies in free space associated with these lengths are 30–3 THz. The human egg itself is around 100 μm, so again the wavelength in free space is 3 THz.

We now turn our attention to the volume energy densities of biological systems compared with free space. It is claimed that THz frequencies are needed to break the DNA bond. But these cellular structures reside in biological systems and not in free space. The volume densities within biological systems given by Eq. (1.1f) can be far greater than that of free space. The energy density of photons in free space is magnified enormously by the permittivity and permeability of biological tissues. Hence a far lower exogenous field in free space than THz is needed to perturb the DNA from its normal role of replication. In 1986 Morgan used an axisymmetric inhomogeneous lossy dielectric model of the human cranial structure for both an 'adult' and an 'infant' to calculate the interior spatial distribution of the absorbed power density. These computations found significant 'hot spots' at both 1 GHz and 3 GHz.

There is a very interesting cosmological analogue of this situation. It appears that the energy densities in the early universe were far higher than they are today. In the 1920s Hubble found the distance between galaxies was red-shifted; the further the galaxy, the bigger the red shift. This led to the big bang theory and its relationship to general relativity discovered by Friedmann and Lemaître. The 20th century saw an explosion of cosmological knowledge in general. For instance, it was found that electrons and protons recombined to form hydrogen atoms when the universe was roughly 380,000 years old, a red shift of $z = 1{,}100$. At this time photons are considered to have decoupled from other matter, travelling through the universe without further interaction to appear today as the cosmic microwave background radiation. While these photons got through as a coherent signal, many others did not. Among these are those that travel between galaxies. In recent years there have been reports to indicate that the universal expansion is accelerating. Dark energy is one accepted theory as to the reason for this acceleration. It is also known that if a medium is expanding then the speed of a particle will increase if a collision process is involved during the expansion. This appears to be true for all particles, including bosons such as the photon. Like all matter, the photon appears to have been getting faster as cosmological evolution has proceeded. As discussed in Chapter 3 a red shift also has an important relationship to biological evolution. It has been accepted that following recombination the photon mean free path became larger than the Hubble length and that these intergalactic photons travelled without interaction between galactic masses. Like the electron whose mean free path differs markedly between moving inside atoms and within conductors such as metals the photon mean free path can be very different depending on its physical state. SFT involves a collision mechanism within the atom where two coherent photons transit between the electron and the proton. There are an integer number of collisions each cycle where the mean free path is the Bohr radius, 0.529×10^{-10} m. As we have seen this leads to an estimate for the photon mass where the photon can be modelled as a tiny particle having a hydrogenic structure. It may be then that the photon has been speeding up as evolution proceeded and the energy density within the universe fell.

Returning to a biological system, as we have seen, it is a red herring to say that THz fields are required to 'break the DNA bond'. Much lower energies can perturb cellular structures, for example, getting the cell to turn on its apoptosis mechanism. Exposure to RF (100–300 kHz) can be beneficial in treating melanomas and brain tumours, but in the same way RF might well be causing harm to growing cells and biological systems.

KEY

1.7 Benefits of EMF Exposure

Over the past three decades there has been a growing consensus amongst medical researchers that while some EMF exposures might be hazardous there are also opportunities that are technological in nature. While the spectrum of EMFs is wide the main thrust of this medical research has centred on the ELF band. The ELF band is of interest because many people live near high-powered transmission systems, including substations, and they use a range of powered appliances in their homes. Epidemiological research has revealed risks associated with H fields generated by ELF sources. Health risks include a wide variety of ills ranging from disruption of normal circadian rhythms to childhood cancers. One important study was that of Wertheimer and Leeper, who examined the use of electrically powered blankets and water beds, finding increased levels of miscarriage, birth defects and low birth weights. These are all problems associated with growth and development.

At the same time as such adverse findings were coming to light, medical researchers were using pulsed EMF (PEMF) exposures either in the form of interstitial electrodes or non-invasive capacitive applicators. Bassett in 1993 noted a range of beneficial bioeffects, including fracture union, spine fusion and osteoporosis, while noting a number of suggested mechanisms, including ion cyclotron resonance, parametric resonance and amplification of signals at the cell membrane. Pilla since 1972 had suggested EMF was a first messenger in a signalling pathway via Calmodulin (CaM) that can induce a chemical cascade. PEMF appears to piggy-back on to one of the body's mechanisms of repair and regrowth. PEMF is effective for bone and wound repair and pain and oedema reduction. As we have seen, an EMF operating at 100–300 kHz can be used to treat melanomas. These fields can cause rotation of the cell nucleus, which is associated with an H field effect. Similarly such cellular rotation if associated with long-term ELF exposures might interfere with the normal cell cycle, causing a malfunction of normal replication consistent with the developmental problems observed by Wertheimer and Leeper. Until we know with some degree of confidence the biogenic processes of development we cannot be assured of the beneficial bioeffects of any particular ELF signals, especially those at the same frequencies as power transmission. It may be that there is a moderate level of a PEMF therapy that is optimal between benefits and adverse effects. One aim of this text is to elucidate the biogenic mechanisms of replication that come together as an SFT balance of the magnetic and electrical energies within the chromosome that occurs at metaphase into anaphase, forming the crux of cellular division. Biogenic fields induce such an SFT balance. Here too a chemical cascade occurs as metaphase gives way to anaphase and the chromatids move to their spindle poles.

The processes may be complex and unknown, but we now know at least what we need to achieve. Biophotonics stands at the cutting edge of a new scientific frontier. The question boils down to whether a range of normal to hypersensitive humans of varying age—young, old, adolescent or adult—undergoing a range of development stages from normal to abnormal can be impinged or assisted by EMF at any frequency and magnitude. What is the benefit or adverse effect for a particular exposure setup? That is one of the aims of the emerging science of medical and therapeutic biophotonics. There are others, including studying the precise details of how the various tissues and organs operate as biophotonic systems. As well as being a biochemical and a genetic system, DNA in its various forms is a biophotonic system that evolves over time. How does this interrelationship play out, and what are the details? We know the body can be both an emitter and a receiver of biophotons. What role does the environment play in this process involving many subsystems? Can the various forms of biophotonic radiation be used medically? These are questions that should reveal the mechanics of emerging life to a greater level of intricacy than previously available when our knowledge was obscured by an incomplete mathematical physics. While we can discern the genetic characters of the book of life, we do not know what the sentences mean, let alone the paragraphs. We stumble upon meaning in a haphazard way at present, sometimes trying, for example, to use chemical means of therapy if we are fortunate enough to see a link between genome and disease. Biophotonics as a form of medical intervention is upon us. We need to study the essential processes that come together to repair, sustain and replicate terrestrial life.

1.8 Mathematics of Endogenous Fields within the Cell Cycle

Over the past 80 years the structures of atoms, molecules and macromolecules have been calculated using QM and other QFTs. These theories are based on Lagrangian formulations in which wave equations are second-order differential equations, compared with the first-order differential equations relating to the E and H fields given by Maxwell's equations. While these potential equations are considered by their waveform to be simpler than Maxwell's equations, because they are second-order they involve complicating gauge functions chosen for particular geometric applications. This integral form of Maxwell's equations results in the photon being effectively modelled as a single-point particle, a form of Dirac delta function, a mathematical singularity in space and time that can induce numerical errors requiring a process called *mass renormalisation* to eliminate singularities. Indeed QFTs are to large degree numerical rather than analytic. Studying the relativistic self-fields of particles in motions Yaghjian in 1998 modelled the electron as a finite sphere of surface charge, revealing serious errors in both the CEM and QM models of the fields as the radius goes to zero. Jackson discussing the validity of the inverse square laws of the classical fields gave their accuracy as 0.1% at macroscopic levels. A more accurate form rather than the field forms of CEM and the quantum methods was at the atomic level.

From 1927 until the recent form of SFT was discovered there was no option other than QM when determining chemical bonds. The implicit assumption was that apart from these numerical solutions, chemistry and EM were complete and nothing more could be learnt either of the photon or of the bonds that kept atoms together and caused them to break apart. Bond lengths and angles of the atoms were the limits of atomic chemistry. Here atomic chemistry equates to the present solutions obtained via QM where the mechanism by which the bonds range between weak and strong is unknown. There was no photon in this mix. 'Hydrogen bond' was 'an area of interest'. Problems of chemical structure and bonding have in fact slowed progress in biology, including the unknown mechanisms of the cell cycle. Using the current implementation of QM it is impossible to understand how metaphase and anaphase follow each other within the cell cycle. Now the reaction can be understood as a cell–cell cooperative effect where chromosomes break into chromatids via photonic processes that are only understood via SFT.

There is a system of photonic structures within atoms and molecules that controls the way molecular bonds range between rigid and elastic forms. Applying SFT to the ML equations for the hydrogen atom, it is seen there are six variables, not four, involved in the complete SFT solution. The range of mathematical analysis now includes how molecules bind together. Varying the energy density, a function of the permittivity ε and permeability μ in a region, alters the eigenvalues of the ML equations that involve not only four scalar equations but also the constitutive relationships in various regions such as the different tissues of the body. Thus

SFT reveals the presence of two previously unknown quantum numbers associated with the photon in its role as the bond within atoms and molecules. These two extra numbers arise as either discrete numbers of photon collisions within atoms or discrete lengths of photon orbits between atoms, electrons and protons being interactions not previously available or known within quantum theory with its HUP and incomplete photon. This collision-based model agrees with classical gas models, providing continuity with the probabilistic studies of Boltzmann and Maxwell.

SFT also sees inside the photon its intrinsic structure. The organisation of the photon into an internal structure is as its extrinsic behaviour also termed *photon chemistry*. Thus photon chemistry is a generalised terminology having wider application than just atomic and molecular EMFs; it also applies to the other various fields in regions where the strong and weak nuclear forces exist, for instance, where photons and phonons can form into gluons. Photon chemistry inside the photon is similar in concept to the well-known atomic chemistry that has evolved over the past 150 years since Mendeleev arranged the elements into a two-dimensional array known as the periodic table. Quantum theory extended this to a four-dimensional array of quantum numbers l, m, n and s relating to the discrete motions of the electrons and protons inside atoms. As well as the photon having a spin similar to the electron, families of photon compounds correlate with the various field particles (bosons) known to particle physics. It may well be that intelligence, memory and other cognitive abilities are biophotonic in nature.

Photon chemistry thus covers a very wide area of physics and biophysics like its two forebears, atomic chemistry and atomic biochemistry. Photon chemistry is indeed a fractal-like image of atomic chemistry. Science has chased the big end of the mass spectrum via colliders and other high-energy engineering; it may be productive to also search the low end of the mass spectrum. There are thus six degrees of freedom associated with EMFs within and amongst atoms (Fig. 1.17). Photon chemistry controls the behaviour of snowflakes, avalanches, magnetic flips of the earth and the sun, the cell cycle and possibly a wide range of physiological and neurological functions, to name some areas of photon chemistry. We sit at the sunrise of a new powerful knowledge.

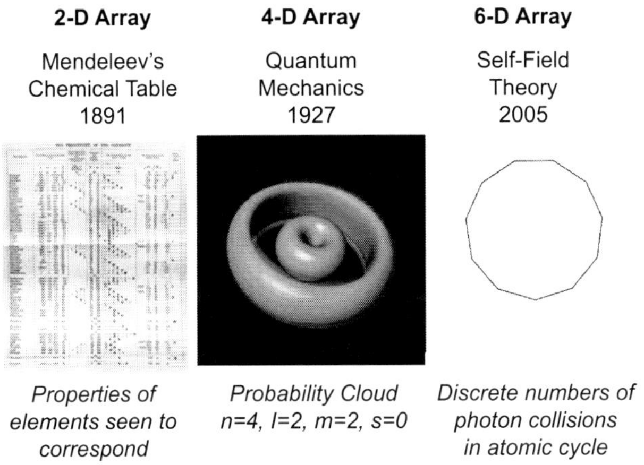

Figure 1.17 Historical perspective of the two new SFT quantum numbers associated with the resonance of the photon in its role as the binding energy in atoms and molecules.

The role of the photon as a binding agent in molecules, including DNA, is a resonance effect, and phase lengths in multiples of $\pi/2$ can be associated with these bond lengths. The chromosome's bond length and its frequency change with the ambient energy density in the cell and the state of the photon. This is a continuous energy process, except when transition energy levels are reached and precipitous changes occur, including breaks of molecular bonds. The quantum nature of these biophysical mechanisms lies in a continuous energy environment and mathematics. The observed flow of photons from strands of DNA is like flows in metals where electrons form into bands within semiconductors.

Though the body can initiate cell replication in many different ways during growth, development, maintenance, regrowth, repair, illness or old age, we shall now focus our deliberations on a 'generic' mitotic cell cycle, in particular metaphase and anaphase. The biological situation is considered to concern either a colony of cells or a tissue consisting of many cells wherein a central cell is located. Observations show the cell and many of its components can stiffen, enabling biogenic EM processes to occur. This stiffening may involve the uptake of water molecules by the cell and/or by the slow evolution of biogenic fields. Cell division occurs as a staged series of intricate dynamic motions with interspersed periods of homeostasis known collectively as the cell cycle. After a period of relative homeostasis and slow development called *interphase* the chromosomes align themselves at the equator of a cell in a stage termed *metaphase* before suddenly cleaving apart at *anaphase*. The cell walls then cleave into two structurally distinct cells, and the cell cycle starts over again. While much is known about mitosis the fine print has not been legible, we have not been able to see the forest for the trees.

Until recently it was unknown how the chromosome split into two chromatids or how the walls cleaved, despite a large body of knowledge of the role of proteins in the division. Gagliardi proposed a process based on electrostatics in the cell, whereby the membrane of the cell nucleus disintegrates and two spindle poles separate to the north and south ends of the cytoplasmic interior of the cell. Dipolar microtubules extend the reach of the intracellular fields to the plasma membrane, forming the cell walls and via the kinetochores to the chromosomes at the centre of the cell. Popp had been conducted experiments looking at emissions of small flows of biophotons from strands of DNA exposed to various forms of EM, including UV light. Although microbiology and biochemistry have revealed a vast quantitative knowledge, for example, of the DNA structure and the human genome, there are many unanswered questions, including the biophotonic mechanisms involved in mitosis.

The salient features of metaphase are the formation of a north–south geometry where the chromosomes align in the neighbourhood of the cell equator in a balance of the expanded and stiffened helical structures. The chromosomes become stretched along the length of the chromatids. As described by Gagliardi microtubules reach out from the kinetochores until they reach the chromosomes from the north and the south. The formation of this balanced structure at the equator of the cell can involve the dielectric properties of a particular tissue consisting of particular cells, one of which may be assumed to be approaching metaphase. Each tissue varies with respect to its extracellular composition as each cell type and its extracellular composition are different. To achieve sufficient energy to rupture the central centromeric bond of the chromosome into two separate chromatids, the surrounding cells become fully polarised. To this end proteins that can diffuse within the plasma membrane of neighbouring cells are slowly polarised as one central cell in metaphase and all surrounding cells align according to the central cell within the polarised population. Each of the diffusing proteins is either charged or dipolar and integral to the membrane by reaching across the membrane, while being able to diffuse within it. Amplifications of the electrostatic fields in this fashion are known from the dielectric constants for various tissues to be up to order of 10^7. The process described above where cell–cell cooperation allows metaphase to continue into anaphase may take minutes or several weeks to evolve. The centromere is ruptured via the slow build-up of electric and magnetic energies within the colony of cells that are all polarised and eventually reach a combined energy that is greater than the bond energy leaving the chromosome with its two sister chromatids no longer structurally sustainable.

This dielectric cell–cell effect is half the overall SFT balance. While the neighbouring north–south-aligned cells act to magnify the original exogenous exposure at the site of the sister chromatids, a second phase of the overall SFT process can provide axially directed fields through the chromatids. These axial fields form a stochastic or noisy magnetic field, causing the membrane proteins in neighbouring cells in the equatorial plane to start rotating, causing an enhanced magnetic field to be fed back to the chromosomes. It is important to realise that this process is a turbulent process like the agitated stirring seen within clouds before a storm breaks. This overall H field causes lowering of the ambient energy and aids in the process of chromosome rupture. There is observed during metaphase a stirring of the cytoplasm. This is due to an H field in the equatorial plane that along with the dielectric multiplicative effect in the north–south direction focuses on the chromosomes that at first stiffen and eventually rupture into chromatids going via metaphase into anaphase. Anaphase is a form of photonic cascade. The photons forming the binding energy of the DNA reach a limit in terms of the photon states going from flaccid to rigid structures.

This photon cascade as applied to metaphase is a description of electric and magnetic forces causing stretching and eventual rupture of chromosomes into chromatids via slow changes to the intracellular energy. Bonds at the centres of the chromosomes oriented in the north–south direction defined by the spindle poles are disrupted. This photonic mechanism may also include sudden flips and changes of the H field due to the structural changes. SFT reveals a new level of detail of the photon in its role as a mediator of EM energy inside atoms and molecules. The neighbouring cells play a role in amplifying the initial polarisation caused by the spindle poles inside the fertilised cell. The combined electrostatic field is applied across the membrane of the original fertilised cell and then relayed along the length of microtubules from the spindle pole to the chromosomes. Metaphase does not continue on to anaphase until the neighbouring cells amplify the original tiny level of extracellular polarisation to a sufficient level, while reducing the required energy level via the magnetic effect in the equatorial plane to rupture the chromosome bonds laying in the north–south direction. This overall process is somewhat similar to the formation of storm clouds to eventually produce rain in an avalanche-type process involving hydration bonds. Clouds become distended volumetrically and puffy, and an agitation, including swirling, can occur before the storm. Like ice or snowflakes that form clouds, a hydration structure envelops the chromosomes. Similarly hydrogen bonds also occur at the central centromeric region.

Fertilisation is the first and most important step used by the body to initiate division where the nucleus is pierced and the nuclear membrane disintegrates, allowing the charged spindle poles to physically separate with the cytoplasm. But there is a range of methods to achieve replication of cells, tissues, blood, etc. This includes protein signals, including heat shock proteins, chemical cascades, endogenous fields and many other techniques used by the body as methods of repair and maintenance. Exogenous fields can also be used to induce changes in the replication process. In the case of the method used by Kirson et al. to treat melanomas the permittivity of the skin at 100 kHz is around 5×10^4, while the permittivity of the cytoplasm has be approximated in previous numerical work at around 80. Hence a magnification of the exogenous exposure results in significant bioeffects on both the plasma membrane during cleavage and a rotation of nuclear material within the cell nucleus.

Geometry plays an important role in cellular mechanics. Théry and Bornens noted that cell rounding is a common feature of cell division where cells can adjust their volume or surface. They also noted stiffening of the plasma membrane cortex and positioning of the spindle poles. Part of this may involve the transmembrane proteins becoming rigid helices. All these mechanisms indicate relatively strong forces in balance. Kleckner et al. found similar findings for chromosomes. In the EM domain SFT balances E and H fields. This may apply to metaphase to form a theoretical description of the stiffening and eventual rupturing of the chromosomes into chromatids in particular bonds oriented in the north-south direction defined by the spindle poles. SFT reveals a new level of detail concerning the photon in its role as a mediator of EM energy inside atoms and molecules. The neighbouring cells, both in the north-south and in the east-west plane, play a role in amplifying the initial north-south currents caused by the spindle poles and the east-west currents within the chromosomes inside the central cell. The combined enhanced E field is applied across the membrane of this initial cell and then relayed along the length of microtubules from the spindle pole to the chromosomes. Metaphase does not continue on to anaphase until the neighbouring cells amplify the original tiny level of extracellular polarisation and magnetisation to a sufficient level to rupture the centromere bonds. The sister chromatids with their chromosome arms have a central region containing heterochromatin and the centromere. Attached to the centromere on both north and south sides is the outer kinetochore that binds to the microtubules north and south. The centromere occupies a central region, north-south, east-west within the chromosome structure that is symmetric relative to both chromatids holding the two sister chromatids together. The central bond in this region is the one eventually broken, allowing the chromatids to be pulled north or south to their spindle poles (Fig. 1.18).

There are many issues concerning human reproduction. A strong correlation exists between industrialisation, fertility and birth rate across the globe, highlighted by a sharp distinction between the underdeveloped and industrialised worlds. The replacement rate of just over two children per family is the exception rather than the

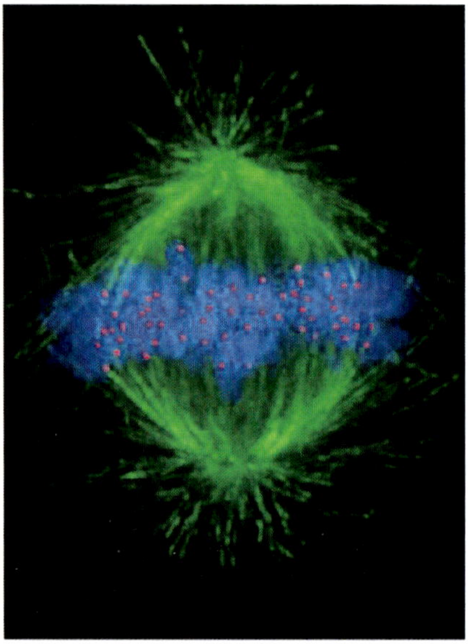

Figure 1.18 Human cell: microtubules in green, chromosomes (DNA) in blue and kinetochores in pink. (Credit: Wikimedia.)

rule in the developed world. Using statistics from Thailand fertility was found lower among women participating in industry than those in agriculture; numbers of pregnancies and births were also lower among industrial workers. Reasons behind such differences are unknown though chemical and technological causes have been considered. It is known industrial and medical technologies using EM heat production increase the risk of spontaneous abortions among workers. The heartache of childless couples in the developed world is contrasted by overpopulated, undeveloped regions at risk of famine and disease. An ethically acceptable and reliable method to manage population dynamics is a much desired outcome of EMF research.

SFT is less than a decade old in its modern form. It extends our knowledge of physics and biophysics to a new level of photonics that applies from below the photon to the size of the cosmos. SFT shows physics, biology and the field are all fractal. Particle physics has to date been predicated on the basis of 'elemental' particles, yet it seems we are living in the middle of an unknown number of levels of interaction. Looking to the future the processes of the cell cycle may be discovered in detail, allowing mankind unprecedented medical abilities that might extend the human lifespan. No doubt there will be revisions of the material in this text. All models of physics and biophysics are just that—models. They do not replace reality only allow predictions of reality by theoretical models. By the time SFT is replaced by a better mathematical physics model, mankind may be living in a world with completely new medical and biological capabilities. The opportunities lie ahead.

Chapter 2

Self-Field Theory

The mathematical history of electric, magnetic and EM fields reveals five distinct periods: pre-mathematical ancient knowledge of fields, electro- and magnetostatic fields, classical EM fields, quantum theories and SFT. SFT is a new description of electromagnetic interactions. At its heart are bi-spinorial motions for both the electromagnetic fields and the interacting particles. Among its recent successes it has solved a simple model of the hydrogen atom, obtained an analytic estimate for the mass of the photon and provided the first glimpse of the structure within the photon. Its eigenstructure goes beyond that of quantum theory, adding arrays of molecules of varying photon states that range in flexibility from solid, liquid crystal, liquid and gas phases. The structure within the photon may yield an organisational structure for bosons reminiscent of the atomic chemical table first noted by Mendeleev in 1860 via a two-dimensional array of elemental properties. The phonon and gluon are involved in this boson organisation. The self-field formulation obtains an analytic expression for Planck's number, providing a basis for its understanding as a variable of motion applying equally to the electron, the proton and the photon. The fields of SFT differ markedly from those of CEM and QFT. The fields in SFT are discrete streams of photons. The photons are specified

Inside the Photon: A Journey to Health
Tony Fleming

ISBN 978-981-4241-40-3 (Hardcover), 978-981-4241-88-5 (eBook)
www.panstanford.com

via a bi-spinorial function as spatially and time-varying motions, including spiral helices between the electron and proton of the hydrogen atom. In SFT, two rotations are used to model EM motions, very different to its forebears. Importantly SFT is fractal, while CEM and SFT are not; SFT is deterministic, while QFT is probabilistic.

2.1 Introduction

The mathematical history of electromagnetics (EM) is intimately linked to the history of mathematics itself. The ancients began to write in mathematical notation almost as soon as they were able to write sounds, including language and music. Such notation was needed for ordinary trade and accountancy purposes. Astronomy was one of the earliest areas of science that demanded a numerical fluency of those charged with the duties of recording astronomical events in order to better predict future trends, including the seasons. It was found, for instance, that crops could be grown in a more effective manner by understanding the seasons. Thus life at the astronomical level was known to be connected to agriculture and the development of life on earth. Yet much of this knowledge was intuitive, and no way of scientifically recording this knowledge existed in ancient times. Thus science was mainly a community-level knowledge passed down by oral tradition, much like religious observance and other areas of community life. Indeed it was not until the mathematics was sophisticated enough that mental concepts of the fields and the photon were able to be widely understood. As seen quantum theory, while recognising the photon, was able to understand neither the photon nor the photonic level existing across science. It turns out that photonic-level interactions, the fields, are responsible for gravitation at the solar system and galactic levels, cosmological and biological evolution and indeed life itself. Humankind needed to attain a certain level of mathematical skill to begin to understand this science.

That the ancients understood both electric (E) and magnetic (H) fields in an intuitive sense is seen in aboriginal tribes and ancient civilisations across the globe. From the aboriginal Africans and Australians, to the first European and Chinese tribes, to the

North and South American Indians, knowledge of one's environment was necessary for survival. Many ancient civilisations, including the Babylonians, Indians and Chinese, used magnetic and electrical energies of a location to help find areas where dwellings could be erected, near water or other geological features on slopes, for instance, or in a particular direction and orientation. Astronomy and magnetic compasses were used to measure direction and are well understood by Western science. At Banpor in western China huts were oriented east–west to utilise the sunlight inside the huts. Most prehistoric aboriginal tribes used similar concepts to organise communities and dwellings. Many of these ancient methods such as feng shui, dowsing, chakras and the third eye, have not been understood nor appreciated by modern science in regard to a number of possible photonic effects. The human body has mechanisms that utilise biophotons that are not currently known or understood using the mathematical theories extant before self-field theory (SFT). That said, recent investigations have begun looking at biophotons that are emitted from the body or absorbed by the body, including the cranium, the hands and the feet. Ancient concepts of yin and yang are similar to concepts of negative and positive electrical charge and dipolar magnetism. Energy balance between yin and yang is a forerunner to SFT.

While ancient civilisations were not as mathematically adept as science is today, we must not presume they did not know of electricity and magnetism at a reasonably deep level. Ancient knowledge is only fractionally known at this point in time; the fires that destroyed the early university and library at Alexandria set science back for perhaps a millennium. It was not until mankind devised the *scientific method* that real progress was again made. During the pre-mathematical ages of ancient times intuition was the main method of scientific research. It is perhaps this aspect of ancient science that supplies some understanding of how ancient science, a communal storehouse of effects and phenomena, became infected by efforts at divination and magic. Modern science is more sceptical and wider known than our ancient forebears, whose rule was mainly by dictatorial decree. We have evolved systems of mathematics and politics to be democratically and cynically inclined. Now we put new theories through a grill of suspicion until they fail or succeed. The ancients had no such method based on intellectual and experimental results, though a few did foresee such a scientific need.

Before Newton and Leibniz both independently invented differential and integral calculus in the 1670s mathematical abilities were restricted mainly to solutions of simultaneous linear and quadratic equations. The discovery of calculus by Newton and Leibniz demanded a new notation, and today we generally use Leibniz's method of indicating a differential and an integral, although Newton's notation is sometimes used as needs dictate. Now the mathematics was in place, the following century provided the other necessary ingredients to enable the mathematical formulation of electrostatics and magnetostatics by Coulomb and Ampere. Advances in mechanical engineering were made, including the torsion spring and the marine chronometer. Coulomb is credited with the invention of the torsion balance that enabled his experiments concerning weak electrostatic charges. A decade or so later Cavendish was able to investigate the gravitation forces between two spherical masses that attracted each other in a horizontal plane.

Like the 1920s, the period just prior to the French Revolution was a period of great change in human affairs. The introduction of Western science to bioelectricity came peremptorily in 1771, when Galvani discovered that the legs of frogs twitched when exposed to sources of electricity. At that time electrical energy was stored in Leyden jars (Fig. 2.1), whose design had been improved by many, including President Franklin of the U.S. While Galvani could replicate his experiments it took 200 more years for these experiments of electric excitation of living nerves to be understood quantitatively at the electronic level.

During this time several important mathematical levels of understanding, of the electrons, ions and finally the fields have been uncovered. Biophotons, the most recent level of knowledge, are basically dipolar particles that interact forming streams of field particles. Their recent mathematical understanding has emerged into the spotlight of current research in a number of areas, including bioelectromagnetics (BEM), where classical electromagnetics (CEM) has been used since the end of World War II (WWII). That photons have a tiny but non-zero mass and therefore weakly attract one

Figure 2.1 A 'battery' of four Leyden jars (credit: Wikimedia).

another is revealed by SFT. We can understand this by examining the unknowns in terms of the number of Maxwell–Lorentz (ML) equations and looking at Planck's number as a measure of the energy within a single rotation inside the hydrogen atom. This reveals a level of interaction within the atom and between atoms not previously known within electrostatics and magnetostatics, CEM or quantum field theory (QFT).

2.2 Electrostatics, Magnetostatics, CEM, QFT & SFT

The history of electrostatics and magnetostatics can be summarised by the equations derived by Coulomb and Ampere, Eqs. (2.1a,b), and their evolution over nearly a century to the Maxwell equations, Eqs. (1.1a–d). The static equations describe uncoupled electric (E) and magnetic (B) fields valid approximations only at macroscopic and microscopic levels.

$$F_{\text{ES}} = \frac{q_1 q_1}{4\pi\varepsilon_0 r^2} \tag{2.1a}$$

$$F_{\text{MS}} = 2\left(\frac{\mu_0}{4\pi}\right)\frac{I_1 I_2}{r} \tag{2.1b}$$

The history of the derivation of the time varying Maxwell's equations includes some early work by Ampere, Gauss and Weber that suggested links between electrodynamics and gravitation in the mid-19th century. Ampere's original equation had an angular relationship specifically contained within it, a forerunner to the angular differences we note between the fields of electrostatics, magnetostatics, CEM, QFT and SFT.

Whatever form of Ampere's equation is used, the original static equations are uncoupled. Only to an uncoupled magnetostatic approximation do the equations of electrodynamics hold for gravitating objects. The result is a spin-less earth model rotating unstably around an equally unstable spin-less sun. The fully coupled form of Maxwell's equations along with the Lorentz equations yields a more complete and stable gravitational model. This model has spin, where each atom gives up one photon to a dipole–dipole interaction that goes beyond current knowledge of atom–atom or molecular interactions. This photon-level interaction between atoms is incomplete within both CEM and QFT. The fractal nature of matter is also connected to this concept of a photon-level interaction, or photon chemistry. In the case of humans and other life forms this interatomic interaction is controllable by choosing to turn their view at will. Biophotons are more ubiquitous than just the atoms in a biological optical system; each photon in a whole-body biofield connects an atom to another atom. In the case of humans we appear able to surprise a person who sees us from an oblique or posterior viewpoint by turning to meet his or her gaze. In general the atom–atom-level interaction becomes a boson-level interaction across all matter within the universe.

In contrast to the uncoupled mathematical form of electrostatics and magnetostatics, Maxwell's equations are coupled E and H fields and are four first-order partial differential equations applying to time-varying phenomena. In their original form Maxwell's equations were 20 in number and the variables were potentials, not the E and H fields of Eqs. (1.1a–d). Maxwell understood these equations as having an underlying wave basis, and he succeeded in deriving second-order wave equations showing the EM phenomenon was a plane-wave propagating in free space at the speed of light. The relativistic Liénard–Wiechert potentials are taken directly between charge points and give the time-varying EM fields for point charges in arbitrary motion relative to one another.

In 1903–1904 Abraham and Lorentz attempted to solve the problem behind what is referred to as the self-field problem, the back reaction of its field on an electron's own motion. They modelled the electron as a spherical surface of current density that moved at relativistic speeds, finding that CEM resulted in a '4/3 problem' where force on the electron contradicts Newton's law $F = ma$:

$$F_{\mathrm{e}} = \frac{4}{3}m_{\mathrm{e}}a - \frac{2}{3}\frac{q^3}{c^3}\dot{a} + \text{'structure terms'}$$

Many tried to modify the classical equations. At that time CEM was seen to be failing. Yaghjian recently reviewed the problem, finding inconsistencies with *both* quantum theory as well as CEM as the radius of the electron approaches zero. He found the generic solution of a charged surface of finite radius moving at relativistic speeds was correct within the macroscopic domain. The self-field problem of CEM theory remained unsolved until recently, when its modern variant succeeded in solving the hydrogen atom. The main reason for this success is the examination of *mutual* effects between *pairs* of particles rather than looking at effects due to single particles.

Einstein by 1915 proposed the equations of both special relativity (SR) and general relativity (GR). His 1905 paper on SR assumed (1) physics is unchanged across inertial frames and (2) the speed of light c is unchanged across inertial frames regardless of whether the light is emitted by a body at rest or by a body in uniform motion. From these postulates he derived the Lorentz transformations and the Fitzgerald–Lorentz time and space dilations. Einstein also showed any ether was superfluous to his deductions. SR was the culmination of efforts to understand why Maxwell's equations were invariant to the Lorentz coordinate transformation. As we have noted relativity is applied as if all dilations are real effects rather than 'hidden' variables. According to SFT relativity involves both the internal and the external motions of the photon. Relativity sees all space as expanding since the big bang. According to SFT this is incorrect and only the external distance must be considered, not the internal or hidden distance. Thus there may be a central point in space-time corresponding to the big bang; there is no reason to assume it has moved. In 1905, Einstein via the photoelectric effect observed that radiation itself, not just its frequency, acted as discrete quanta or particles of energy $E/h\nu$ termed photons. Bohr, recognising the Plank–Einstein equation $E = h\nu$ held for emitted and absorbed energy, proposed a quantum theory of atomic spectroscopy in which angular momenta were whole numbers of Planck's quantum $\hbar = h/2\pi$. Over the next two decades the methods of quantum theory evolved via Bohr's classical theory of the atom as a heuristic eigenvalue method based on examining the Lagrangian associated with a single particle.

The history of QFT began with the evolution of Schrödinger's non-relativistic equation to Dirac's relativistic equation. Schrödinger expressed the phase of a plane wave as a complex factor, Eq. (2.2a). The Klein–Gordon equation, Eq. (2.2b), is a relativistic version of Schrödinger's equation. Using Feynman notation, Eq. (2.2c), we see how the Klein–Gordon equation can be analytically factorised to yield the positive and negative energy versions of the Dirac equation, Eq. (2.2d), where the γ_μ are 4×4 bi-spinors of mathematical nature. Thus Dirac predicted the electron's anti-particle, the positron. Although Dirac's equation is first-order it is derived from a second-order wave equation, and it is this that demands gauge symmetry and invariance under conformal transformations.

$$i\hbar\frac{\partial}{\partial t}(x, t) = \widehat{\mathrm{H}}\Psi(x, t) \tag{2.2a}$$

$$\left[\nabla^2 + \frac{m^2c^2}{\hbar^2}\right]\psi(r) = 0 \tag{2.2b}$$

$$\left(\partial^2 + m^2\right)\psi = 0 \rightarrow (i\partial + m)(i\partial - m)\psi = 0 \tag{2.2c}$$

$$\left(\gamma_\mu\partial_\mu + im\right)\left(\gamma_\mu\partial_\mu - im\right)\psi = 0 \tag{2.2d}$$

In SFT the ML equations, Eqs. (1.1a–d), can be algebraically rewritten in matrix form as in Eq. (1.5) where each particle has two spinor motions σ_i^{EM} and a current κ_j^{EM}.

$$\mathrm{M}_{ij}^{\mathrm{EM}}\sigma_i^{\mathrm{EM}} = \kappa_j^{\mathrm{EM}} \tag{2.3}$$

The motion of the electron is an exact EM self-field solution, and its position at any time is the sum of two spinors forming a bi-spinorial motion $\sigma_{\mathrm{o}}^{\mathrm{EM}}(r_{\mathrm{o}}, \omega_{\mathrm{o}})$ and $\sigma_{\mathrm{c}}^{\mathrm{EM}}(r_{\mathrm{c}}, \omega_{\mathrm{c}})$, where the distance of the electron is written as a sum of the spinors, not a Pythagorean root mean square but a function of orthogonal orbital and cyclotron spinors:

$$r^{\mathrm{EM}}\left(r_{\mathrm{o}}^{\mathrm{EM}}, \omega_{\mathrm{o}}^{\mathrm{EM}}, r_{\mathrm{c}}^{\mathrm{EM}}, \omega_{\mathrm{c}}^{\mathrm{EM}}\right) = r_{\mathrm{o}}^{\mathrm{EM}} e^{j\omega_{\mathrm{o}}t} + r_{\mathrm{c}}^{\mathrm{EM}} e^{j\omega_{\mathrm{c}}t} \tag{2.4}$$

If the intrinsic energy of the system changes from that of free-space ε_{o} and μ_{o} the EM fields within the atom can adjust, altering the atomic and molecular binding structures. In effect this means that the radial and spin states of the two photons involved in the binding energy can adapt to the energy change where the photons themselves are assumed to have a composite structure. Now the system has six degrees of freedom, including the E and H fields, in response to the change in ambient energy. The two extra variables give a range of variation orthogonal to the phase diagrams of atoms. A typical phase diagram shows only a solid line separating the various phases. SFT indicates a small range of binding structures that depend on the intrinsic energy of the system.

The photon streams of SFT are taken between centres of rotation, not between charges as in CEM and QFT. Thus two distances are involved in the bi-spinorial motion and not one. The fields in SFT are discrete quanta, photons, rather than the continuous vector fields of CEM. QFT also models the fields as discrete quanta. However, each individual field in QFT may be taken over the entire range of solid angles connecting any number of particles, depending on the geometry and number of interacting atomic particles. The field can be shown as a small wavy line within Feynman diagrams (see Fig. 2.2), but the QFT mathematics does not specify any actual path, only the start and end points where Dirac-delta functions are used to insert propagator kernels and Greens functions, depending on the field and its geometry. In QFT the field is actually a potential, not

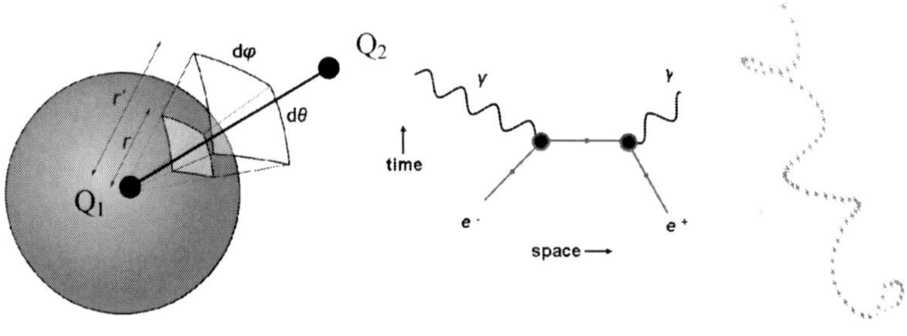

Figure 2.2 Field forms. (a) CEM, (b) QFT and (c) SFT.

the E and H fields of SFT. In QFT the fields (potentials) are modelled as impulse functions specified at charge points. Within SFT, a pair of particles defines the bi-spinorial field and its motion, this pair of particles and its bi-spinorial field form a unique couple. Due to orthogonality the fields of this couple do not influence any other charges apart from the couple. The transit of the SFT field is specified via the bi-spinorial function and assumes various motions, including spiral helices as it transits between the electron and proton within the hydrogen atom. There is a vast difference between the SFT time-variant field motion and the time-invariant CEM where the field ubiquitously covers all solid angles with no definition other than its vector nature as to the actual field motion, field flux being the only indicator of field motion. Similarly the uncertainty of the field within QFT is related to its lack of a complete and coupled EM bi-spinorial field form.

There are other major differences, including an absence of Heisenberg's uncertainty principle (HUP) within SFT. As the photon is modelled via bi-spinors uncertainty is obviated. In SFT the electron's self-fields are modelled via a complete EM function that explicitly includes both E and H fields, enabling the complete analysis of the mutual self-field effect between two particles. Unlike the quantum potentials that are expectations yielding probabilistic solutions, the bi-spinorial field variables of SFT allow completely deterministic solutions. This results in a clearer picture of the physics that includes the particle–photon interactions and the binding mechanism. The solution is complete (coupled) and based on the first-order ML equations; hence neither SR nor gauge symmetry is problematic. The bi-spinor field variables of SFT are a priori relativistic and its solutions analytic rather than numerical. In EM applications gauge symmetry is maintained by analysing pairs of particles and not just single particles. Similarly a pair of conjugate photons of finite mass do not constitute *'symmetry breaking'* of the Lagrangian as in QFT.

2.3 Strong Nuclear Fields

One important aspect of SFT and the bi-spinors that represent the EM motions of the electron, the proton and the photon is the hypothesis that there is an associated relationship based on tri-spinors that represent the strong nuclear motions of quarks and gluons inside the proton. As is well known, there are three quarks inside the proton. In this case there is a self-field interaction between triplets of quarks where the interaction sub-matrix is of size 3 $\times$ 6 per particle, as distinct from the EM field sub-matrix that is of size 2 $\times$ 4. Now each particle has *three* spinors that mutually link the three quarks together. Each spinor operates in a plane, and the three planes form an orthogonal set in three-dimensional (3D) space. This means that there exists a system of Maxwellian equations having additional differential components such as to form four scalar equations instead of the three associated with Maxwell's EM equations.

Like the EM interaction the strong nuclear fields controlling the motions of charged particles satisfy the following adaptation of the ML equations. In general, the region is assumed isotropic and homogeneous, and ε_{n}, μ_{n} and μ_{n} are invariant scalars. Where nuclear sub-particles, quarks, carrying units of elementary charge q_q are studied, the modified ML equations can be written

$$\nabla \bullet \vec{\mathrm{E}} = \frac{q_q}{v_q} \tag{2.5a}$$

$$\nabla \bullet \vec{\mathrm{H}} = 0 \tag{2.5b}$$

$$\nabla \bullet \vec{\mathrm{N}} = 0 \tag{2.5c}$$

$$\nabla \times \vec{\mathrm{E}} + \mu_{\mathrm{n}} \frac{\partial \vec{\mathrm{H}}}{\mathrm{d}t} = 0 \tag{2.5d}$$

$$\nabla \times \vec{\mathrm{H}} - \varepsilon_{\mathrm{n}} \frac{\partial \vec{\mathrm{E}}}{\mathrm{d}t} = \frac{\pi}{s_q} q_q \vec{\mathrm{v}} \tag{2.5e}$$

$$\nabla \times \vec{\mathrm{N}} + \nu_{\mathrm{n}} \frac{\partial \vec{\mathrm{E}}}{\mathrm{d}t} = 0 \tag{2.5f}$$

where the modified Lorentz equation for the forces acting on the quarks is

$$\vec{\mathrm{F}} = q_q \vec{\mathrm{E}} + q_q \vec{\mathrm{v}} \times \vec{\mathrm{B}} + q_q \vec{\mathrm{v}} \times \vec{\mathrm{M}} \tag{2.5g}$$

and constitutive equations $\vec{\mathrm{B}} = \mu_{\mathrm{n}} \vec{\mathrm{H}}$, $\vec{\mathrm{D}} = \varepsilon_{\mathrm{n}} \vec{\mathrm{E}}$ and $\vec{\mathrm{M}} = \nu_{\mathrm{n}} \vec{\mathrm{N}}$ where ε_{n} μ_{n} and ε_{n} are invariant scalars, the nuclear constitutive parameters similar to those of free space, ε_0 and μ_0, except the energy density within the nucleus now depends upon the three gluon fields

$$\mathrm{d}U_{\mathrm{N}} = \rho_{\mathrm{N}} \mathrm{d}V = \frac{1}{2}(\varepsilon_0 \tilde{E} \cdot \tilde{E} + \mu_0 \tilde{H} \cdot \tilde{H} + \nu_0 \tilde{N} \cdot \tilde{N}) \mathrm{d}V \tag{2.5h}$$

where $\tilde{N}$ is termed the nuclear field and $\tilde{M}$ is the nuclear flux density. As with $c = (\varepsilon_0 \mu_0)^{-1/2}$ there are corresponding relationships between the gluon speed and the ratios of the three fields. These modified equations provide three orthogonal motions per quark. There are 6 unknowns per particle, 18 in the nuclear system. The curl equations, Eqs. (2.5d–f), provide four scalar equations, and there are two virial equations to give six equations in six unknowns. Like the photon, the resulting analytic parametric solutions may be compared to the experimental results given by particle physics.

In the strong nuclear case there are 3 × 4 scalar curl equations plus 3 × 2 force balance equations, making 18 equations in 3 particles altogether that can be solved to yield 6 quantum numbers per particle in nuclear structures. This compares with the 2 × 3 scalar equations from Maxwell's curl equations plus 2 force balance equations, totalling 8 equations per 2 particles, giving 4 quantum numbers per particle for atomic structures. The two extra quantum numbers agree with the experimental observations of high-energy physics. The three quarks have three-way streams so that each particle performs spinor motions in three planes and not two as in the case of EM-related particles such as the proton and the electron.

Now of course the only way that the quarks can perform these additional spinor motions is if the photons themselves have formed into compounds *beyond* 'ordinary' or low-energy photons. When photons are inside high-energy regions they may exist in higher-energy states and compounds. Like atomic structures then, photons may *also* form into structured compounds. In the case of quarks, the photon particles may have a third spinor motion when they form into triplets. The forces these photonic compounds carry yield the third component of spatial direction and they *only* carry this extra force component *inside* the nucleus. In examining all possible candidates for the identity of this third field and its associated force, the only reasonable candidate turns out to be the acoustic (A) field and its associated pressure force.

Although mitosis is an EM phenomenon and rain clouds are a nuclear phenomenon, the phenomenon of mitosis is very similar to the formation of lightning, wind and thunder from cloud formation within low-pressure regions. According to SFT, storm clouds can be treated as large molecular aggregates. As the temperature and pressure vary with height, the atomic and nuclear binding energies within the aggregate may vary and changes in E, H and A fields occur. It is possible that variations in quantum states of both EM and nuclear binding energies drive the production of lightning, wind and thunder within the cloud. The aggregates thus change their energy state, giving off E and H fields as photons and A fields as phonons. The energy associated with the E field is emitted as a translational energy and drives the production of lightning. The H field is emitted as a rotational energy that drives the production of wind. The longitudinal energy is released as acoustic energy that produces thunder. Time variations in both lightning and wind may also induce a transformation of E and H fields into A fields and vice versa. Thus it may not only be the spatial variations of height, pressure and temperature but also be the time variations of energies associated with the cloud masses. These cloud masses are just large molecular aggregates that undergo energy state changes that are emitted or transformed into various forms of lightning, thunder and wind (Fig. 2.3). This is similar to the energy changes used in spectroscopy or the beats from a stethoscope.

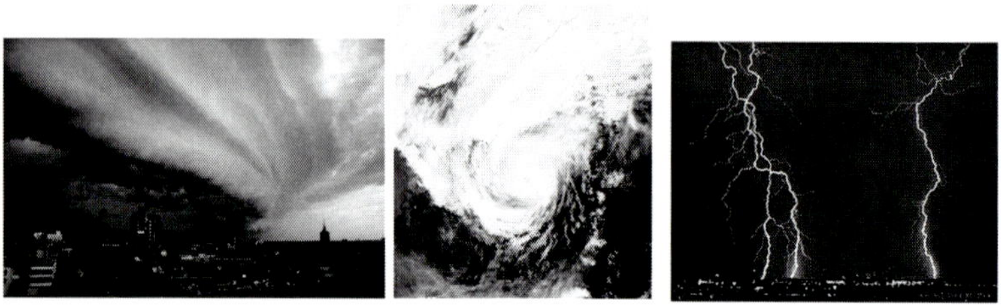

Figure 2.3 According to SFT, lightning, wind and thunder are forms of electric, magnetic and acoustic energy (*source*: www.stormscapedarwin.com).

2.4 Connectivity and Biophotons

The Einstein–Podolsky–Rosen (EPR) paradox lies at the heart of the way SFT sees physics and biophysics. Einstein co-authored a paper on the paradox that challenged quantum theory he saw as incomplete. Quantum entanglement is now understood as the way in which parts of a quantum mechanical system are connected. The quantum states of the constituent parts are linked; one part cannot be properly described without mention of all other parts within the system. Einstein referred to this as 'spooky action at a distance'. SFT too sees quantum theory as incomplete. SFT suggests a missing coordinate within the photon that is the underlying reason for HUP and its lack of knowledge concerning photons. Like entanglement SFT suggests that photons link all atoms together as in a crystal lattice. In solid crystals according to SFT the atoms rotate in synchronised fashion. Photon interactions apply between virtually all atoms within the universe in which dipole–dipole EM interatomic forces exist as streams of photons between masses. Deoxyribonucleic acid (DNA) is an example of a liquid crystal capable of 'stiffening' as the biophotonic states change within the cell cycle. Biophotons link to objects around a biological system. This includes the biological ability of optical sight but can include other biophotons within the complete body field. Biophotons are found in specialised physiological tissues, including elasmobranch fish, which use jelly-filled canals called the ampullae of Lorenzini on the lower peripheries of their fins for predation. As well as being absorbed, biophotons are emitted by biological systems.

Einstein was right; it is now realised that an interconnection existing between living and non-living systems has been overlooked, while quantum theory failed to see the photonic- and biophotonic-level interaction.

In their normal role biophotons convey energy and information to parts of a biological system, in addition to their role as a binding energy at the EM level. One form of photonic information is phase length, whereby biological systems can infer positions and lengths of their surroundings for various purposes, including sight, touch, smell, motion, location and predation. A body's fields connect it with its environment. These whole-body fields or biophotons are no different to ordinary connective photons except that they perform this informative role. Within the body biophotons can perform this informative role and in addition provide the distribution of energy to its cells, organs and tissues. DNA is central to this role of supplying energy during the cell cycle. SFT suggests that self-organisation within the cell occurs because of the reducing level of energy as the cycle proceeds. Biophotons provide both the information and the energy when a group of cells decides to replicate their own species.

2.5 Biophotonic States: Liquid Crystal to Liquid and Back Again

In biological terms perhaps the most important finding due to SFT is a quantitative understanding of the transitional chemical changes from a liquid state to a liquid crystal state that occur to macromolecular arrays as observed in the various structural forms of DNA. These chemical changes are directly related to biophotonic binding energies and states. Well known to atomic physics is the shell structure of the outer electrons in multi-electron atoms. In the original Bohr theory, a construction rule for shells was a maximum of two electrons per orbital in order of increasing orbital energy. In 1936 Madelung came up with a better matching set of rules. Orbitals are filled in order of increasing quantum numbers $n + l$, where orbitals have the same value of $n + l$ and are filled in order of increasing n. This gives the following order for filling orbitals: $1s$, $2s$, $2p$, $3s$, $3p$, $4s$, $3d$, $4p$, $5s$, $4d$, $5p$, $6s$, $4f$, $5d$, $6p$, $7s$, $5f$, $6d$ and $7p$. In

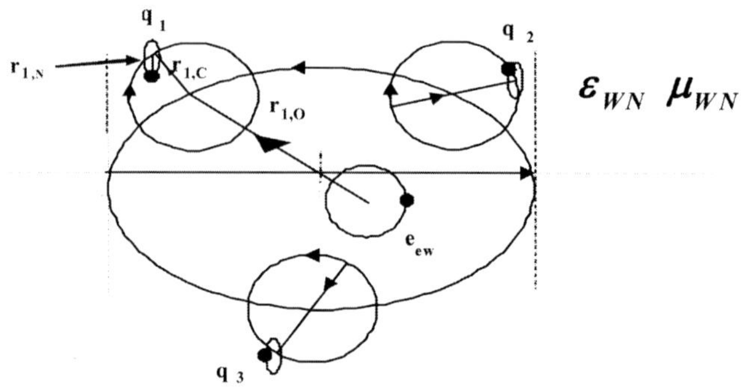

Figure 2.4 Electroweak and nuclear shell structures. Analogous to the EM electron shifting shell as energy density is raised via $\varepsilon_0\ \mu_0$, so too the weak electron moves its inner shell if $\varepsilon_{WN}\ \mu_{WN}$ is raised; at the same time the proton shifts its nuclear shell.

nuclear physics a modified form of the atomic shell model is used to describe a set of nuclear shells in terms of energy levels.

SFT reveals another shell structure for the weak nuclear electrons and their complementary proton shell structures within the nucleus. As shown in Fig. 2.4 an internal SFT balance can be achieved between a weak electron and a proton within the central core or inner spherical region of the nucleus. In this region elevated energy densities are present via the electroweak constitutive parameters $\varepsilon_{WN}\ \mu_{WN}$. What this implies is that the strong nuclear effect occurs within an annular region where the quarks and gluons move.

In the early days of quantum mechanics (QM) Pauli suggested that the chemical bond was due to a resonance phenomenon. We can see that despite his insight, and his brilliance in using QM to obtain a wealth of chemical information, the QM method itself was incapable of determining the exact bond relationships due to HUP and the lack of a complete EM field form. The history of both QM and atomic chemistry is organic, with numerical and other approximations used to aid the progress in atomic chemistry that was then blossoming. That there were errors in QM was not considered, although Condon pointed out the limitations, including singularities and numerical intractability, that were becoming apparent. EPR pointed to the incomplete nature of quantum theory.

Atomic and molecular bonds are described by using orbitals, orbital diagrams and electron configurations. As discussed the orbitals are filled in order of the lowest potential energy for the atom. The maximum number of electrons in each orbital is two according to the Pauli exclusion principle. According to Hund's rule, orbitals of identical energy are considered half-filled with parallel spin before they are paired. Such procedures are approximations used, in part, to avoid the need for quantum calculations where QM is numerical and intractable for many-electron problems. These recipes and their QM numerical counterparts have grown somewhat organically since the first days of chemistry as properties of various molecular substances have been studied via numerical calculations.

The Pauli exclusion principle goes back to the early days of chemistry, when it was realised that atoms and molecules with even numbers of electrons were generally much more stable than those with odd numbers. In 1919 Langmuir suggested that the electrons might be clustered in some fashion explaining features of the periodic table. Groups of electrons were thus thought to be gathered into shells. Bohr modified his theory to account for special numbers of electrons $(2, 8, 18, 32, \ldots)\ 2n^2$ that corresponded to 'closed' shells. Pauli following discussions with Bohr realised that the number of electrons could be defined using four quantum numbers. For this purpose he proposed a new two-valued quantum number that was later identified as electron spin. For the electron $s = \frac{1}{2}$ gives the total angular momentum of the electron as $S = \sqrt{\frac{1}{2}\left(\frac{1}{2}+1\right)}\hbar$ corresponding to the two possible spin states, 'spin up' or 'spin down', $S_Z = \pm\frac{1}{2}\hbar$ for the electron.

In SFT terms the same principle relates to the way that electrons and protons orbit as EM binaries in a stable double rotation. If a second pair consisting of another paired electron and proton is displaced in phase by π radians, as shown in Fig. 3.9, the combined motion and structure are also stable. Underpinning this is a physical principle in the way the photons cannot stream between more than two charges particles at the sub-atomic level simultaneously in a stable fashion. While photons can transit between a particular electron and its paired proton as a pair of streams, they cannot move

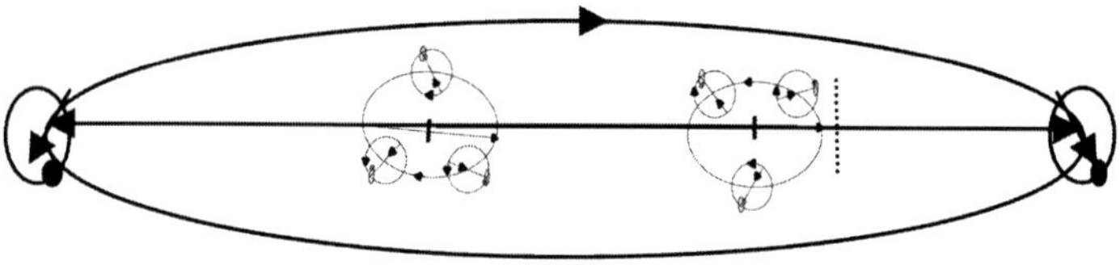

Figure 2.5 Pauli exclusion principle. Hydrogen molecule modelled as two protons with a composite quark structure and two electrons.

(orthogonally) between any other electrons or protons. This is a fundamental realisation that greatly simplifies the determination of atomic and molecular structures. Thus the four EM particles shown in Fig. 2.5 are connected by only two (pairs of) photon streams, two E field streams and two H field streams that must rotate in synchronous fashion with the electrons and protons.

This synchronous motion also applies to multi-atom configurations, including solid crystal arrays. The modern version of the double-slit experiment can be easily understood where the two slits are made of the same solid piece of material. If the two slits are synchronous versions of each other then the slow build-up of the final diffracted image photon by photon is no longer enigmatic. Hence arrays of matter appear to be synchronous in their motions in the same way as shown in Fig. 2.5.

Now the difference between a solid and a range of less rigid arrays of atoms can be understood via the resonance of the binding photons. As these binding photons change their resonant phase length over integer multiples of $\pi/2$, the whole array becomes less and less rigid until the array can resemble a liquid. This process can continue until the whole array becomes disassociated and each atom is independent in a gaseous phase.

A short summary of the mathematical and basic physical findings of SFT is as follows. On the basis of the motion of the photon within the hydrogen atom SFT can be used to obtain an analytic expression for m_γ, the mass of the photon, and provide a theoretical rationale for α, the fine-structure constant. Recently, SFT was used to derive actual motions of the electron and proton within the hydrogen atom in the form of eigensolutions to a system of partial differential equations based on the ML equations. The atom was modelled mathematically via two point-mass particles, the electron and the proton, an extension to the Bohr model. Planck's number $\hbar = \frac{q^2}{4\pi \varepsilon_0 v_e}$ is found to be the energy per cycle of the principal eigenstate that depends on the motions of the electron, proton and photon, all involved in the dynamic balance of the atom. The photon performs many relativistic transitions back and forth between the proton and the electron within each cycle of the electron and the proton that rotate coherently about their centre of mass. The phase length of the photon each time it transits $\pi/2$ maintains the overall coherency of the atom's periodicity, providing a method for analytically comparing the energy of the photon with that of the electron, $m_\gamma c^2 = \frac{\hbar \omega_\gamma v_e}{4c}$, where ω_γ is the collision frequency of the photon. Assuming some non-linear polygonal motion circumscribes a circle representing the Bohr mageton, the photon collision frequency is estimated as 54 using the known value of the fine-structure constant $\alpha = \frac{v_e}{c} = \frac{4m_\gamma c^2}{\hbar \omega_\gamma}$ and the Landé g-factor Thus m_γ evaluates to 0.396×10^{-55} kg (0.221×10^{-19} eV). Since the photon must perform a discrete number of transits per cycle this suggests collisional-based polygonal rotations for both the electron and the proton rather than the assumed circular rotations given by spinor theory used by both SFT and QFT. This suggests a way to check more detailed SFT models of the hydrogen atom that include a structured nucleus. Inside the hydrogen atom the photon substructure introduces no new physics beyond that given by QM. The role of the photon is, however, much more refined, and another level of mechanism is seen to involve a discrete and regular number of elastic collisions between the photon and the electron, and the photon and the proton. Within molecules, however, the photon substructure introduces two previously unknown quantum numbers by which molecular bonds can range between strong and weak structures. This mechanism is behind the long-range dipole–dipole forces known as gravitation. This photon mechanism appears involved in a number of energy-/temperature-dependent molecular processes. The cell cycle and other biological processes may well depend on the various hydration structures found within and around DNA and other biological protein structures.

Physical phenomena show dependence on hydration structures as in avalanches and the formation of cloud layers within the ionosphere or on the innate energy levels of bosons such as the magnetic flips of the sun and earth.

Chapter 3

Biological and Cosmological Evolution

Galen, a brilliant surgeon of the AD 2nd century, studied his craft by treating sports injuries in his hometown of Pergamon in the Roman empire. He learned many eclectic fields of knowledge, including astronomy at Pergamon library, second only to the one at Alexandria. Moving to Rome he treated gladiators at the Coliseum before becoming physician to the emperor Marcus Aurelius. Anatomy has long been a basis for medicine, but we have recently begun to learn how the body and perhaps the mind are constructed from astronomical and cosmological forces. Biological species evolved forming a phylogenetic 'tree of life' in a specific temporal order ranging from single-celled organisms to microbes, to algal mats, to dinosaurs and to man, including fractal and synergistic relationships between life forms. Obtained via self-field theory (SFT), a new mathematical theory of the gravitational structure for our galaxy may help explain a recently discovered biodiversity cycle of 62 million years (Ma). Rather than survival of the fittest, the 'small picture' seen at the terrestrial level, the 'big picture' of biodiversity shows dependence on blackbody radiation of EM fields within the solar system and acoustic (A) fields within the galaxy. SFT suggests the gravitational structure within the cosmos and cosmological evolution are related to the evolution of physiological structures

Inside the Photon: A Journey to Health
Tony Fleming

ISBN 978-981-4241-40-3 (Hardcover), 978-981-4241-88-5 (eBook)
www.panstanford.com

and biological evolution on the earth. These links reveal a wider evolutionary process whereby the universe has been expanding after the big bang, evolving into superclusters, galaxies and solar systems. Cosmological evolution relates to the energy spectrum of both EM and A field radiation available on the earth in the past over time and the size and structure of species as they evolved. Genera may have evolved in situ from ~3–4 billion years ago after the earth–moon system formed when biophotonic and acoustic spectra on the earth were more energetic and life forms smaller and less solid than now. Life may be evolving across the universe. Intelligence, memory and other cognition may link to the evolution of the cosmos and its gravitational structure.

3.1 Introduction

The mathematical description of physics and biophysics given by self-field theory (SFT) includes exact solutions for the motions of molecules and macromolecules such as deoxyribonucleic acid (DNA). This molecular description involves the eigenstructure of the Maxwell–Lorentz (ML) equations having six degrees of freedom and extends to gravitational arrays of molecules. At the solar system level gravitation is a di-electromagnetic effect involving both dielectric and diamagnetic effects. This includes Newtonian gravitation plus the spins of the planets. Other unique forms of SFT gravitation apply to the various structures observed within the cosmos, including galaxies and higher domains of structure.

The gravitational model of the universe shown in Fig. 3.1 is based on the first four levels of the gravitational structure given by SFT and the four levels observed to hold across the visible universe. Although the motions are simplified to circular orbits, the actual orbits can be less symmetric, such as ellipses, or other more general motions. The galactic orbit can have three rotations. Each form of gravitational motion and its mathematical structure is a differential form of a specific system of ML equations having four, six or more equations. The orbit is a toroid at the next level higher than the galaxy. While Newton's form of gravitation holds across the universe, it is the first term in a series of forces as we go from the solar system

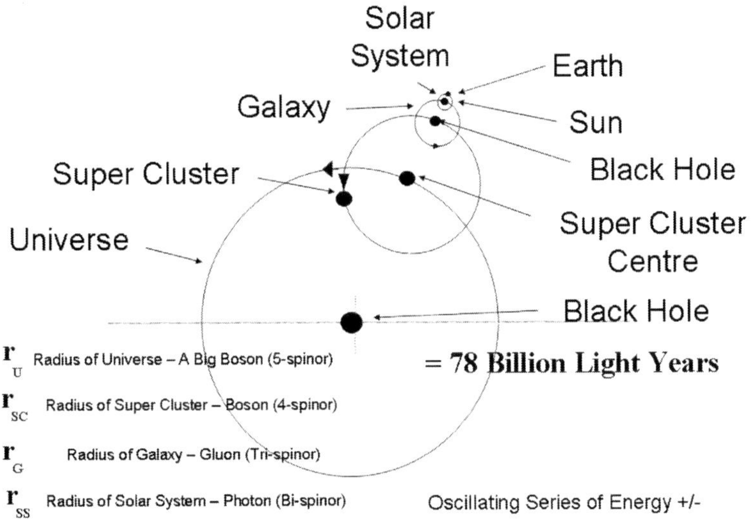

Figure 3.1 SFT model of the gravitational structure of the universe.

up via our galaxy. The model is a combination of particle physics, cosmology and SFT. There are thus distinct levels of gravitation given by SFT, each a balance between matter and fields of different types. The four levels depicted are 'Universe', 'Supercluster', 'Galaxy' and 'Solar System'. Each field of higher, smaller energy in the fractal sequence is a type of integral of a larger but lower energy field in which it is imbedded. Each mass in a lower domain is a differential form of a mass of a higher domain. These differentials and integrals are linked to the analytic forms of the SFT spinors. Thus stronger but smaller domains and weaker but larger domains of forces act across the universe. Alternating positive and negative collations of energy (black and white holes) move near the centres at each domain. Conservation of energy means the universe has a zero sum total energy (creation ex nihilo). This relates to the negative spin of magnetic fields, which is associated with lower temperatures. Hence conservation is also fractal ranging from the universe, to particles and to fields. All matter receives and emits field quanta. There is a link between the number of rotations and the type of gravitation. This may include rotations greater than three. The model presents a kernel mathematics for the unification of physics, a '*standard model*' of SFT.

$$\begin{aligned}
W_{\mathrm{U}} &= S_{\mathrm{U}} + V_{\mathrm{U}} + K_{\mathrm{U}} + B_{\mathrm{U}} \\
S_{\mathrm{U}} &= W_{\mathrm{SC}} = S_{\mathrm{SC}} + V_{\mathrm{SC}} + K_{\mathrm{SC}} + B_{\mathrm{SC}} \\
S_{\mathrm{SC}} &= W_{\mathrm{G}} = S_{\mathrm{G}} + V_{\mathrm{G}} + K_{\mathrm{G}} + B_{\mathrm{G}} \\
S_{\mathrm{G}} &= W_{\mathrm{SS}} = S_{\mathrm{SS}} + V_{\mathrm{SS}} + K_{\mathrm{SS}} + B_{\mathrm{SS}} \\
S_{\mathrm{SS}} &= V_{\mathrm{SS}} + K_{\mathrm{SS}} + B_{\mathrm{SS}} \\
V_{\mathrm{SS}} &= -G\left(\frac{m_S^{\mathrm{SS}} m_{\mathrm{Me}}^{\mathrm{SS}}}{r_{\mathrm{Me}}} + \frac{m_S^{\mathrm{SS}} m_V^{\mathrm{SS}}}{r_V} + \frac{m_S^{\mathrm{SS}} m_E^{\mathrm{SS}}}{r_E} + \ldots\right) \\
K_{\mathrm{SS}} &= \frac{1}{2}\left(m_{\mathrm{Me}}^{\mathrm{SS}} v_{\mathrm{Me}}^2 + m_V^{\mathrm{SS}} v_V^2 + m_E^{\mathrm{SS}} v_E^2 + \ldots\right) \\
B_{\mathrm{SS}} &= f_{\mathrm{SS}}\left(m_\gamma^{\mathrm{SS}} c^2\right)
\end{aligned} \tag{3.1}$$

The model sees energy as extrinsic or intrinsic within each domain. Two forms of extrinsic energy exist, potential energy V and kinetic energy K; W is the total energy, B the binding energy and S the self-field energy of the particles. The system is simplified with only one sub-particle at each level. The kinetic and potential energies maintain a balance. As all motions are evolving (expanding) the balance is inexact. While the orbits are illustrated as simple circles, they are irregular, piecewise or chaotic depending on the actual dynamics—on the fields and their ability to suddenly transit states. This includes the case where magnetic poles '*flip*' according to the photonic energies involved. Since photonic energies are continuous there may be dramatic and sudden flips as photonic transition energies are crossed and the photon states change polarity. The whole cosmological structure involves an expansion process evolving over time, the boundary value problem of the universe.

Fields and matter in SFT are fractal so too are the life forms. Human life is based on the various levels of biological organisation, including the organs, tissues, cells, proteins and DNA. Various microbial life forms exist in the human gut. All life evolved in a time sequence that depended on the associated cosmological evolutionary process. This cosmological evolution involves the observed gravitational structures; the evolutionary process in general occurred as a 'top-down' sequence of events involving higher energies first and lower energies later. We see that the solar system

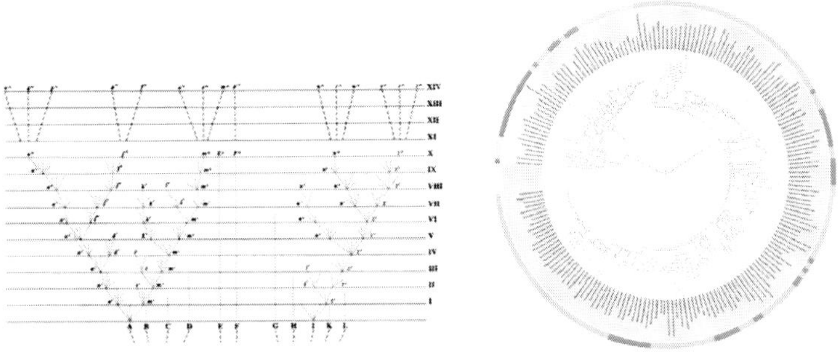

Figure 3.2 (a) Tree of life from Darwin's *On the Origin of Species by Natural Selection* (1859). (b) Tree of life based on sequenced genomes.

level was the last domain to evolve. In the brief initial inflationary stage of cosmological evolution the various bosons, the *photon chemistry*, to support the gravitational structures were evolving top-down, and dependent on both, the various life forms also evolved. The evolutionary concepts of SFT are wider than those of Darwin and genetics, which is based on atomic chemistry (see Fig. 3.2a,b).

While quantum theory allowed a stunningly successful period of mathematical physics and biophysics compared with what preceded it, it was, as Einstein had insisted, incomplete. Ironically, Einstein's general relativity was also incomplete for similar mathematical reasons. Einstein might have been aware of this in his long, fruitless search for unification across physics in his later years. Perhaps the most fundamental assumption on which quantum theory, atomic chemistry and general relativity all depend is the belief there are universal 'constants', including Newton's gravitational 'constant', Planck's 'constant', the electron and proton masses, the speed of light and the fine-structure 'constant', all thought constant across the universe. Many, including Planck, thought this to be the case resulting from the cosmological principle of Einstein's general relativity. According to SFT, these 'constants' vary with the energy density in which they are measured. Even the electron and proton masses depend on the self-energy of their internal structures and the ambient energy density in which they exist, and this varies across solar systems and galaxies since there are gravitational domains across the universe, for example, galactic lenses. If we are to communicate with intelligent beings from exoplanets, it will likely not be via universal 'constants'. Further if evolution varies with energy density in an expanding universe then it is unlikely that evolution on exoplanets either in our own galaxy or in other galaxies is at an equal stage to the earth. The earth began biological evolution 4.5 billion years (Ba) ago, and the universe is 13.7 Ba old; so beings on exoplanets may have evolved ahead of us. Biological evolution creates for each being across the universe a digital fingerprint more differentiated via photonics than previously realised. Biological structure depends on the cosmos and the variations in energy within space-time.

3.2 Gravitational Structure within the Universe

Historically a widespread and considered opinion existed from about 1687 till the 1930s that gravitation throughout all levels of interaction within the universe was similar to that in the solar system, as proposed mathematically by Newton. Lagrange in the late 1700s followed the mechanical and astronomical paths in Newton's footsteps, confirming and extending the application of Newton's gravitational laws to the full extent of the then known universe. Newtonian gravitation is an inverse square law of separation, hence a function in one variable. This was a time pre-quantum physics and pre-relativity during the 18th century, when the universe was thought of as a nebula, the term 'galaxy' did not appear in any index and 'cosmogony' was used in discussing how the solar system formed out of the universe. In 1915 Einstein extended Newton's inverse square law to a system of geodesic equations that took into account the speed of light treated as an observable constant; this introduced a metric equation. In effect it was an extension to a second degree of freedom within gravitation at the level of the solar system.

In the 1920's Hubble used the 100-inch telescope at Mount Wilson to discover that the local 'star groups' were in fact other galaxies. He subsequently discovered cosmological red shift, a discovery that led eventually to the theory of the big bang. In 1931 Oort observed the red shift of stars near the galactic plane (GP). He found that the variation of orbital speeds with radial distance away from the galactic centre (GC) were fairly invariant unlike the solar system, where orbital speeds diminish with radial distance. Zwicky in 1932 extended this finding by studying galaxies within the Coma Cluster measuring similar results. If gravitation were Newtonian, the stars should fly apart; the Milky Way galaxy and its myriad counterparts within the universe were thus now problematic in terms of their gravitational structure. Over the intervening years to the present, one hypothesis is that dark matter in the form of massively compact halo objects, or weakly interacting massive particles, amongst other ideas supplies the missing matter. Another hypothesis is MOdified Newtonian dynamics that heuristically changes Newton's constant of gravitation to match the observed rotational speeds.

As we have seen SFT depicts Einstein's general relativity as the first approximation to a series of forces that forms the gravitational structures in the universe. This series is composed of spinorial terms for the fields that begin with the electromagnetic (EM) field having two interacting spinors and then the gluonic field having three interacting spinors. The gravitating masses also have spinorial motions. At the general solar system level the masses have bi-spinorial motions, while at the galactic level the masses have tri-spinorial motions. Both the field particles and the mass particles possess internal mass and energy. This system is fractal, existing at different discrete levels or domains corresponding to biological systems.

Recently a separate and associated line of inquiry was opened up when cycles of biodiversity were found to exist within the fossil record. Rohde and Muller in 2005 found cycles at 62 Ma. Medvedev and Melott confirmed the cycles in 2007 and discussed them in terms of the solar system's oscillatory motions orthogonal to the GP. There are further reasons linked to the extinction of large, structured mammals to support the finding of cycles of biodiversity within a tri-rotational system of dynamics, which we shall discuss in section 3.2. Galactic and solar system resonances of acoustic (A) and EM energy correspond with the evidence, including cycles, within the fossil record. The orbital speeds of galaxies also correspond with the presence of the A fields that Hans Jenny demonstrated were linked to the formation of solid structures. This galactic tri-rotational dynamics can be linked to SFT via a modified system of ML equations to describe regions where the strong nuclear (SN) interactions occur. The normal ML equations become modified—Maxwell's equations usually consist of two curl and two divergence equations, four in total—and become six by adding an extra curl and an extra divergence equation. These six equations support a system of quarks and gluons within the nucleus that have a three-way interaction between triplets of particles and rotate in three orthogonal directions, this tri-rotational motion linking SFT to the recent analysis of the galactic tide by Klacka. This is a SFT tri-spinorial form of gravitational structure.

3.2.1 *Boson Streams*

The well-known Einstein–Podolsky–Rosen (EPR) paradox lies at the heart of the way SFT sees gravitational physics at the solar system level, which is described as a dipolar interaction between atoms within their near- or far-field zones of interaction where far-field can include astronomical unit (AU) (the mean distance between the earth and the sun) and larger distances. Einstein challenged quantum theory, which he saw as incomplete, in a paper that contained the paradox. As a consequence of this historical debate *quantum entanglement* is now understood as a way in which all parts of a quantum mechanics (QM) system are connected. The quantum states of the constituent parts are linked together so that one part cannot be completely described without mention of all other parts within the system. Einstein referred to this as 'spooky action at a distance'. SFT too sees quantum theory as incomplete. SFT suggests a missing coordinate within the photon is the underlying reason for Heisenberg's uncertainty principle and its incomplete knowledge at the photonic level. According to SFT photons bind all atoms and molecules together, the reason for Einstein's spooky action at a distance. This photon interaction applying to all matter within the universe is not treated within quantum theory.

SFT sees two previously undiscovered quantum numbers linked to the discretisation of the bispinorial motions of the electron and proton within the atom that are supported by the photon in its role as binding energy. This applies to crystals where atoms rotate in synchronised fashion and to biological systems in which DNAs are liquid crystals, where biophotons link to other objects in the view of the biological system. While this implies a biological capacity for optical sight it can encompass other forms of biophotonic emission and detection. This includes the complete body fields of biological systems in general but is found in specialised physiological tissues, such as the elasmobranch fish, which use pores, the ampullae of Lorenzini, on the periphery of their dorsal fins for predation purposes. Human perception is more than just visual and may be neurological. The human brain may be able to sense the biophotons of other conscious beings—not telepathy but a sensory ability to physically synchronise with other body fields. It may be that like eyes that can act as a directional antennae, so too the cranium may be able to swivel around like a biophotonic directional antenna system. This is part of a new understanding we might glean from the photon in its role as a binding energy. So not just the sciences associated with atomic chemistry but also previously unknown roles whereby the biophoton connects living systems. Body fields are involved in this perception of other biophotons around us. Similarly we also send biophotonic signals.

At the level of the macromolecule, and below, the physics of this photon-level interaction is a resonance phenomenon as is the case inside atoms, molecules, liquid crystals or crystals where it acts as a binding energy and its phase length ranges over multiples of $\pi/2$. The interaction can also be a long-range dipole–dipole or gravitation effect whereby the molecules in a massive conglomerate are linked to other nearby or remotely located massive conglomerates. The phase lengths involved are now averaged across a range of different wavelengths where the individual atomic masses are probably not in dynamic equilibrium, depending on the dynamic state of the gravitational system involved. Nevertheless the ensemble average will form a dynamic equilibrium and be a multiple of $\pi/2$. Photons and the derivatives of their flux densities, including their radial derivatives, act as a binding energy between masses. At the atomic level there are relatively few streams of photons, the streams act only between pairs of atoms and the photons move in double rotations as per the general solution of partial differential equations.

At the level of the solar system the streams couple all conglomerates via an interconnected array of forces. The photons emitted and absorbed by the atoms within the conglomerates still stream between atoms taken two at a time. Photons as distinct from gluonic bosons, for example, cannot physically form streams between more than two atoms at any one time. Gluonic structures on the other hand form streams between triplets of quarks, depending on the energy density available inside nuclei. Gluons can also form long-range or gravitational forces between remotely located quarks residing in different massive conglomerates. Individual quarks interact only in triplets; conglomerates at the macroscopic level interact with other conglomerates taken in triplets.

Photons acting as a binding energy between gravitational masses within the solar system, such as between the sun and the earth, lose and regain their energy in their periodic transits into deeper regions of space. Similarly when the strong force gluons exist outside the nuclei they lose and then regain some of their energy during transit through deep space, converting into lower-energy fields that are carried as a binding energy within the galaxy. When they reach their destination the energy rises until they reach the particular nucleus they connect to when they again have sufficient energy to interact via elastic collision. It appears from SFT that bosonic collisions are involved in this gravitational role. This role of bosons as a binding energy between massive particles is the underlying reason for the existence of streams of field particles noted to exist at the solar system level as used by the National Aeronautics Space Administration (NASA). Physically these streams may exist in deep space, waiting to be discovered as regions of relatively higher energy when compared with normal deep space. If a probe should enter one of these streams it may find the energy density rises beyond that of ordinary space where no such streams exist. In fact a similar system of streams has been used by Lo working at the Jet Propulsion Laboratory at the California Institute of Technology and NASA to design a low-energy transit for the Genesis mission flight path. The streams have been described as 'a "freeway" through the solar system resembling a vast array of virtual winding tunnels and conduits around the Sun and planets . . . can slash the amount of fuel needed for future space missions.' Such streams are used as staging points for interplanetary travel. Similar streams composed of bosons exist at all levels of gravitational structure.

3.2.2 *Bosons and Gravitational Structure*

Binney and Tremaine in 1987 approximated the galactic tide within the GP as a function of x and y with no gravitational effect due to the z direction using a Cartesian system at the GC. Klacka in 2009 derived an equation of motion for the galactic tide that included the z component. Thus there are three components of galactic gravitation distinct from the two components of gravitation of the solar system. These three components of gravitation link directly to SFT where the SN forces involve three degrees of freedom. Fleming has recently discussed the relationship between gravitation and the differential forms of ML equations. While the photon has an internal structure consisting of two internal particles having equal mass, the gluon can by analogy be described as a three-particle internal structure. At the EM level the ML equations are written in matrix form, where each particle has two spinor motions σ_i^{EM} and a current κ_j^{EM}.

$$\mathrm{M}_{ij}^{\mathrm{EM}}\sigma_i^{\mathrm{EM}} = \kappa_j^{\mathrm{EM}} \tag{3.2}$$

The motion of the electron is an exact EM self-field solution, its position at any time is the sum of two spinors forming a bispinorial motion $\sigma_{\mathrm{o}}^{\mathrm{EM}}(r_{\mathrm{o}}, \omega_{\mathrm{o}})$ and $\sigma_{\mathrm{c}}^{\mathrm{EM}}(r_{\mathrm{c}}, \omega_{\mathrm{c}})$ where the distance of the electron is written as a sum of the spinors not a Pythagorean root mean square but a function of orthogonal orbital and cyclotron spinors: The electron's motion has four unknowns, as shown in Fig. 3.3:

$$r^{\mathrm{EM}}\left(r_{\mathrm{o}}^{\mathrm{EM}}, \omega_{\mathrm{o}}^{\mathrm{EM}}, r_{\mathrm{c}}^{\mathrm{EM}}, \omega_{\mathrm{c}}^{\mathrm{EM}}\right) = r_{\mathrm{o}}^{\mathrm{EM}} e^{j\omega_{\mathrm{o}}t} + r_{\mathrm{c}}^{\mathrm{EM}} e^{j\omega_{\mathrm{c}}t} \tag{3.3}$$

The SN formulation can be written using Eq. (2.6) in compact form, where each quark has trispinorial motions σ_i^{SN} and currents κ_j^{SN}.

$$\mathrm{M}_{ij}^{\mathrm{SN}}\sigma_i^{\mathrm{SN}} = \kappa_j^{\mathrm{SN}} \tag{3.4}$$

The quark motions may be an exact SN self-field solution, depending on the energy densities, their position, the sum of three spinors forming a tri-spinorial motion $\sigma_{\mathrm{o}}^{\mathrm{SN}}(r_{\mathrm{o}}, \omega_{\mathrm{o}})$, $\sigma_{\mathrm{c}}^{\mathrm{SN}}(r_{\mathrm{c}}, \omega_{\mathrm{c}})$ and $\sigma_{\mathrm{t}}^{\mathrm{SN}}(r_{\mathrm{t}}, \omega_{\mathrm{t}})$. In this case the total distance of the quark is written as

Figure 3.3 Plot of a bi-spinorial solution for electron motion in $r_o - \varphi_o$ and $r_c - \theta_c$ planes, where $r_o = 4r_c$ and $\omega_c = 16\omega_o$.

a sum of three spinors again, not a Pythagorean root mean square but a function of orthogonal spinors, one orbital, one cyclotron and a third designated tangential. These equations may provide a closed-form system of nuclear solutions.

$$r^{SN}\left(r_o^{SN}, \omega_o^{SN}, r_c^{SN}, \omega_c^{SN}, r_t^{SN}, \omega_t^{SN}\right) = r_o^{SN} e^{j\omega_o^{SN} t} + r_c^{SN} e^{j\omega_c^{SN} t} + r_t^{SN} e^{j\omega_t^{SN} t} \quad (3.5)$$

Differential forms of the EM (DEM) and SN (DSN) ML matrices exist where the rotations take place in orthogonal planes to their undifferentiated SFT forms. This includes the axial spins of a planet or solar system compared to the cyclotron motions of an electron or quark:

$$\mathrm{M}_{ij}^{\mathrm{DEM}} \sigma_i^{\mathrm{DEM}} = \kappa_j^{\mathrm{DEM}} \quad (3.6)$$

$$\mathrm{M}_{ij}^{\mathrm{DSN}} \sigma_i^{\mathrm{DSN}} = \kappa_j^{\mathrm{DSN}} \quad (3.7)$$

According to SFT galactic gravitation is related to a differential form of the SN ML equations: The internal three-particle structure of the gluon, and the tri-rotational dynamics of quarks within the atomic nucleus, corresponds to the tri-rotational gravitational dynamics (compare Figs. 3.4 and 3.5). Looking at the general picture of SFT in terms of the bosonic structures involved, and observing the evidence from cosmology, the universe is segmented into layers of structure that depend on the spatial variation of energy density to form solar systems, galaxies, superclusters and perhaps the universe itself. According to SFT each layer has its own gravitational structure, depending on the energy density and the bosons available as a function of space-time. Biological systems also consist of layered structures, including tissues, organs, cells, chromosomes, neural networks and biophotonic pathways.

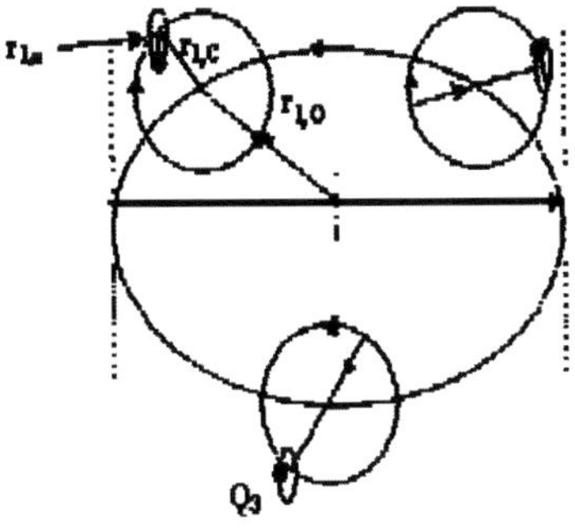

Figure 3.4 Model of a proton consisting of three quarks, each having three degrees of motion.

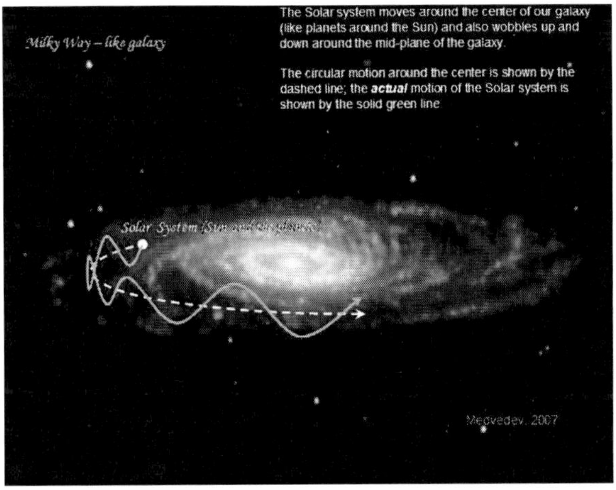

Figure 3.5 Galactic dynamics. A typical sun orbiting around the centre of a galaxy (credit: Medvedev and Melott).

3.3 Biodiversity May Be a Resonance Process

The general meaning of biodiversity is the variation of life across all levels of organisation. Recently there have been findings of cycles within domains of biodiversity. In 2005 Rohde and Muller performed a statistical analysis of Sepkoski's compendium on the stratigraphic appearance of 36,380 marine species, sorting the genera into the 295 time periods used by Sepkoski. They produced a

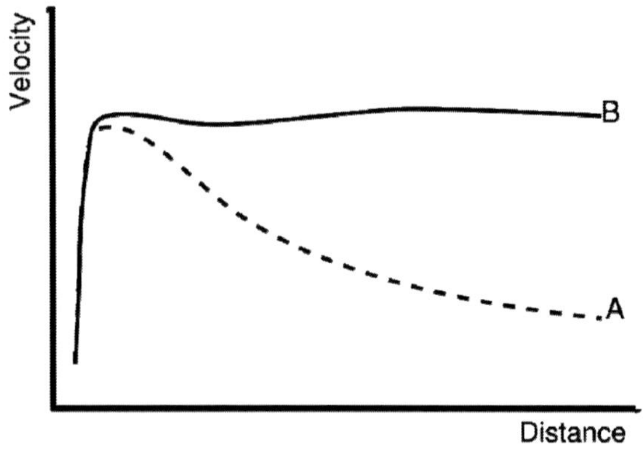

Figure 3.6 Rotational velocity from the centre of the galaxy.

plot of diversity against time and found a strong Fourier component at 62 Ma. They suggested seven possible reasons for the cycles but were themselves unconvinced by any of these hypotheses, only suggesting that such strong statistical evidence should be explicable. In 2007 Medvedev and Melott re-examined the initial data and confirmed the findings of a biodiversity cycle of 62 Ma, suggesting that cosmic ray extinction events may be responsible as the solar system bobs up and down. At its maxima the earth is less protected from nearby cosmic ray events. While cosmic ray extinction events may be more prevalent at the maximum amplitudes of the earth's galactic motion, there are other possible factors of biodiversity.

If the energy density at the earth's surface were constant, this would imply an infinite and possibly homogeneous region surrounding the solar system and the earth. In fact the galaxy is observed to resemble something between an ellipsoid and a flat disk. Unlike the solar system there is depth and structure orthogonal to the GP. Also the galaxy has a reasonably constant orbital speed unlike the orbital speed relationship found within the solar system (Fig. 3.6). The galaxy therefore rotates approximately like a solid. Acoustic fields are known to be involved in structural integrity. Hans Jenny (1904–1972) studied the interaction of acoustics on the formation of structure within particulate substances such as sand, liquid and iron filings, examining the resonant acoustic modes on circular and square sheets of metal. As the earth and the solar system move up and down orthogonal to the GP, the E, H and A fields on the earth's surface vary. The galaxy may be considered to be represented by an ellipsoidal galactic structure within which particles move representing the solar system and the black hole at the GC. The E, H and A fields on the earth's surface vary with distance above and below the GP. This cyclic variation is partly due to the falling-off in energy density near the upper and lower edges of the galaxy as the mass density changes abruptly to that of intergalactic space. The variation in the fields may be involved in the observed cycles of biodiversity.

3.3.1 *Galactic Dynamics*

SFT suggests that the strong nuclear fields controlling the motions of nuclear particles in atoms acting gravitationally at the galactic level satisfy the tri-spinorial mathematics given by Eq. (3.7). These equations provide three orthogonal rotations per quark. Where the photon and the phonon are seen to be composite in structure, each having two sub-particles, it is reasonable to propose the nuclear field is acoustic. Referring to Fig. 3.1, boson streams connect galactic suns to the black hole at the centre of the Milky Way galaxy. Black holes in SFT are seen as analogous in some ways to suns. Suns both emit and radiate energy. They are centres of EM mass, where planets orbit around these central masses. They have positive energy emitted as heat. But they also have nuclear energy, which they share with the centre of a galaxy. This energy may be the cause of solar winds. Similarly black holes are central mass objects, one level of gravitation up from suns; they both emit and transmit nuclear radiation; they are centres of nuclear mass, where solar systems orbit around them.

Thus there is a long range, three-way interaction between strong nuclear masses taken as triplets. This gravitational structure is an extension to that proposed for gravitation at the solar system level where EM dipole–dipole forces are involved and bi-spinors apply between pairs of masses, as in Newton's and Einstein's gravitational models. In the case of the galaxy it is the addition of the longitudinal or acoustic field to the EM forces that adds the solidity to the galactic structure that is observed via the particularly high orbital speeds observed within galaxies. This form of gravitational structure is discussed in a parallel session at this same PIERS conference. The main features in terms of biodiversity cycles are a rotation in the (R, z) plane, where $R = \sqrt{x^2 + y^2}$, and the presence of a photon–phonon gluonic field involving E, H and A fields. Like the streams of photons, the binding energy within atoms, there will be streams of gluonic bosons acting as the binding energy within the galaxy. The density of these streams and the energy within the galaxy itself will fall off near the upper and lower edges of the galaxy. If the solar system's motion orthogonal to the GP approaches these upper and lower edges then the fields will vary at these points in its motion. Like the great pressures at the bottom of the sea the earth moves from the GP towards regions of much reduced pressure at the edges of the galaxy.

3.3.2 *The Role of Acoustic and EM Resonance in Biodiversity*

As the earth bobs up and down to its maxima in its motion within the solar system around the GC, it moves to regions where the acoustic fields are diminished and the structural integrity of matter on the earth is reduced, limiting the ability of species to maintain their physiological integrity. Physiological integrity refers in part to the mechanisms a body uses to reach a state of constancy, homeostasis and adaptation and includes perfusion, nutrition and oxygenation. Tissues that were solid at or near the GP will be more viscous or more solidified and may not have the required structural integrity for life to be as viable at the maxima points in the galactic motion. There is a link between fluid dynamics and acoustics that is pertinent to the current discussion. As is known the speed of sound is given by $c_s = \sqrt{\kappa/\rho_0}$ where ρ_0 is the mean density and κ is the bulk modulus measuring a substance's resistance to compression. Similar equations hold for gases, liquids and solids, as well as biological tissues. If the bulk modulus of tissues falls below a certain threshold the tissue will lose its structural integrity. This is similar to the observation of space sickness that astronauts enduring prolonged stays in free space can suffer deleterious effects to bone, muscle, perfusion and blood pressure. This might also be the case for large structures such as the dinosaurs that roamed the earth some 62 Ma ago, a time of near maximum excursion form the GP.

Similar to the variation in acoustic fields near the upper and lower edges of the galaxy, the E and H fields are influenced by the solar system and its evolution within an expanding universe. At the galactic domain, the black hole near the centre of the galaxy is a negative source of energy unlike the sun near the centre of the solar system, which is a positive source of energy. SFT indicates a fractal view of the universe and its gravitational structures. This includes an oscillating series of gravitational forces that goes beyond the galactic level. Energy density in the bulk medium within the universe is observably convergent rather than divergent. Hence as the solar system bobs up and down, the relative energy due to the galaxy is reduced and the solar system may get slightly warmer than while moving near the GP. Reproductive processes, including mitosis within species, may become marginally less viable as the frequencies of the energy required during the cell cycle are modified away from their values within a more 'habitable zone'. 'Goldilocks zones' are thought necessary to sustain life both at the solar system and the galactic levels. Popp has investigated the spectral emissions from the DNA of numerous life forms, finding wavelengths from 200 nm to 800 nm. At room temperature $T = 290$ K the thermal radiation of Planck's black body is a maximum at $\lambda = 10$ mm, the mean size of human cells. At the sun's surface temperature, 5,840 K, the maximum wavelength is 497 nm in the middle of the band found by Popp (Fig. 3.7). While these lengths and temperatures are interesting more study is needed to see if they have evolutionary significance.

Since the end of WWII, spectroscopy at radio frequencies of interstellar space has revealed the unexpected presence of organic molecules, including formaldehyde (H_2CO) and the amino acid glycine (NH_2CH_2COOH). That such biomolecular structures were found in interstellar space indicates the possible widespread nature of life throughout our galaxy and the universe (see Appendix C).

In 1952 Miller and Urey conducted an experiment to test if amino acids, the building blocks of life, could be generated from simple compounds, water (H_2O), methane (CH_4), ammonia (NH_3) and hydrogen (H_2), as shown in Fig. 3.8. The liquid water was

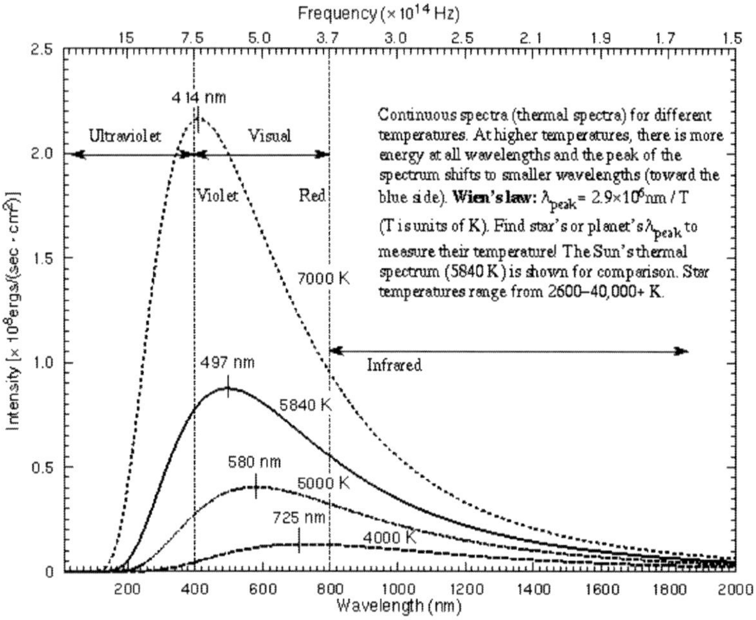

Figure 3.7 Planck's black body (credit: http://www.astronomynotes.com/light/s4.htm).

evaporated by heating and sparks generated across electrodes to simulate lightning. The system was now cooled, causing the water to condense and trickle back, forming a continuous cycle. After 24 hours the mixture had turned pink in colour, and after one week 10–15% of the carbon within the system was organic compounds; 2% of the carbon had formed amino acids involved in the process of making proteins in cells, with glycine as the most abundant. Sugars and liquids also formed. Nucleic acids were not formed within the reaction. But 20 amino acids were formed in various concentrations. Subsequent experiments showed that pre-biotic amino acid synthesis could occur in hydrothermal vents and by delivery to the earth via meteorites (see http://en.wikipedia.org/wiki/Miller%E2%80%93Urey_experiment).

While spectroscopy reveals amino acids are present within interstellar space, these experiments demonstrate how environmental levels of EM energy could also have been involved in the formation

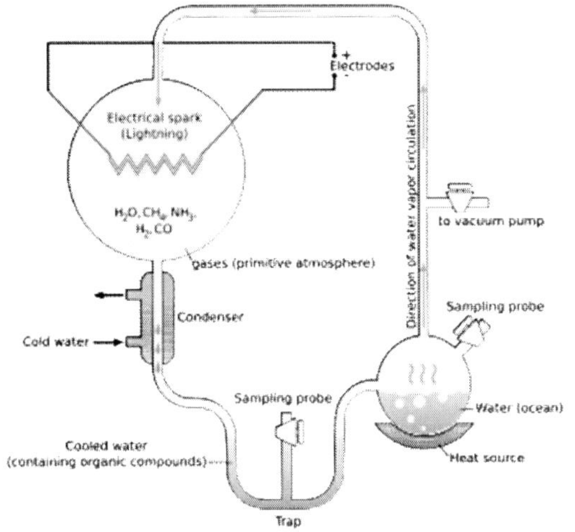

Figure 3.8 Miller–Urey experiment, 1952. Production of amino acids using conditions thought to exist on the proto-earth (credit: Wikimedia).

of pre-biotic amino acids. Acoustic energy could also have been involved as atmospheric or submarine pressures, which would have been higher than today due to cosmic expansion. Photonics and phononics are therefore seen to be central to the origins of life on earth. But there are other early-life associations between SFT with its photon chemistry and the big bang with its subsequent universal expansion. What were the conditions when strands of DNA formed, and how did they form and why? The double helix shape of DNA is suggestive of an SFT process involving amino acids acting like spinors, perhaps being stretched elastically by strong, vertical gravitational forces within the surface of the proto-earth where energy cycles due to a violent and chaotic weather system and strong cyclic gravitational forces were in place. At some point chemical reactions turning EM energy into chemical and photonic forms of stored energy may have been the driver behind the formation of strands of elastic molecules that could act to take currents across boundary regions of heterogeneity within the proto-earth's hydrated surface regions such as rock pools, lakes and lagoons with their sand and other weathered particles. Again, how did primitive cells form, and what where the conditions on the earth at this time? The earliest form of the cell might have been a bubble of inhomogeneity, where ions diffused across a structured membrane that separated an inner spherical region from an external region. This spherical region might have adapted to energy cycles driven by the proto–solar system. Evolution is seen as a process of differentiation. Rather than a chaotic struggle for existence, sunlight and galactic energy are differentiated into evolving life forms, where cooperation between cells provided an important evolutionary advantage.

At the same time, we know the universe has expanded over the past ~3–4 Ba so that the temperature on the surface of the earth has become lower, less chaotic and more stable. Sequentially we can surmise that proto-DNA would have evolved first followed by the proto-cell due their relative size. Universal expansion has meant that one year after the big bang, around 13.7 Ba ago, one light year has stretched to 1,100 light years. This may sound counterintuitive, but it must be remembered that the universe is much larger than 13.7 billion light years, being perhaps 78 billion light years in radius. In a period of superluminal inflation, as Guth suggested in 1980, the universe expanded by a factor of more than 10^{28} to something close to its present size in the first tiny instant after the big bang. Whenever the DNA and the cell evolved, temperatures on the earth were thus much higher than today. It seems from their sizes and the fractal and synergistic nature between them, DNA evolved before the cell. Photonics and phononics performed a central role in the evolution of the universe and the earth and life on earth and across the universe.

3.4 Cosmological and Biological Evolution May Be Linked

SFT sees a link between the observed cycles of biodiversity and the galactic gravitational structure. This hypothesis provides a new form of gravitation applicable to galaxies different to solar systems that includes both photons and phonons. According to SFT photons and phonons react to form gluons in regions where the energy density is sufficiently high. The ML equations can be modified to give a mathematics applicable to strong nuclear regions. Consequently, solidification can occur in galactic structures, corresponding to findings of fairly constant galactic orbital speed variation with a radial distance known since the 1930s. Universal expansion follows the work of Hubble in the 1920s in identifying a red shift applying to matter within the universe. Biological evolution may therefore correlate with cosmological evolution. It is known that the first simple forms of life around 4 Ba ago in the case of our own Milky Way galaxy had single-celled structures. It was around this point in cosmological time that the solar system is thought to have first formed. The small size of early proto-life forms and their bulk modulus relative to mammals corresponds to the fact that the energy density of photons and phonons within the early Milky Way galaxy would have been relatively higher than during later epochs, when mammals of varying sizes evolved. Life forms would be expected to grow larger and be more solidified via phonon resonance within the evolving galaxy as the photon energy density within the solar system. Correspondences exist between the inflationary process of cosmological evolution into gravitational structures and their life forms.

Theories of galactic evolution form a complex array including top-down hypotheses where large gas clouds condensed and bottom-up hypotheses where coalescence occurred of smaller progenitor structures. Many include dark-matter predictions not supported by SFT but rather a modified form of gravitational structure based on a strong nuclear form of Maxwellian equations that includes solidifying phonons within the structure. Observations of the early universe suggest a bottom-up process, where an early galaxy was composed mainly of gas with fewer stars. The gas at a critical point in time underwent rapid contraction, causing the galaxy to rotate faster until it became a very thin, rapidly rotating disk. This is a changing galactic structure, whereby it increases in size and mass until condensation occurs. What causes or ends the contraction is not known, nor are current theories successful at predicting the observed rotation speeds, sizes and masses of disk galaxies.

In the same way the structure of a galaxy evolves, so too does the structure of the life forms it sustains (Figs. 3.9 and 3.10). The three-dimensional forces holding a galaxy together create the pressure

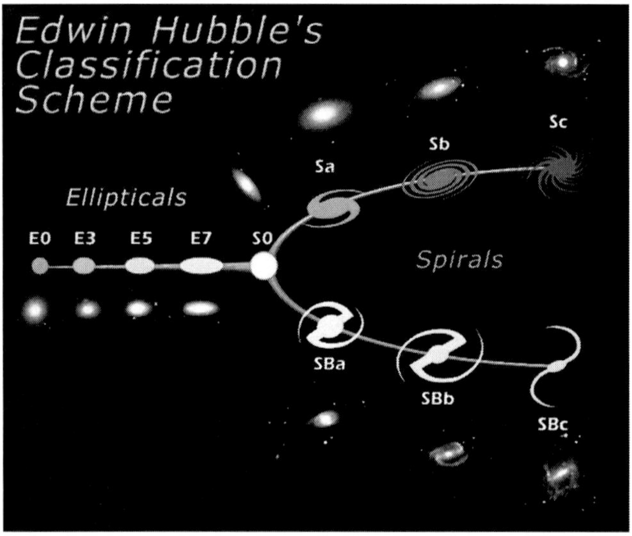

Figure 3.9 Classification of the galactic structure. 'Some are simple, while others are very complex in structure. As one of the first steps towards a coherent theory of galaxy evolution, the American astronomer Edwin Hubble, developed a classification scheme of galaxies in 1926. Although this scheme, also known as the Hubble tuning fork diagram, is now considered somewhat too simple, the basic ideas still hold.' (Credit: Wikipedia.)

within the gaseous, liquid and solid matter inside the galaxy. As this three-dimensional character evolves, so too the gravitational forces evolve within the galaxy. This concerns both the photon and phonon energy densities as a function of the shape of the galaxy. As the galaxy grows, the height in the plane orthogonal to the GP grows. This increase in galactic size and height gives more three-dimensional solidification to the life forms it can sustain. If a species can use this added solidity to its advantage then there is no reason why it will not evolve to a more solid entity. So too as the big bang expansion process continues there is a lowering of the photon energy within the universe and a lowering of the EM energy density and a consequent increase in wavelength available via EM radiation. Hence growth of life forms can expand in a similar way to the solidification process. We would expect to find that biological evolution follows a path towards more solid and larger species as

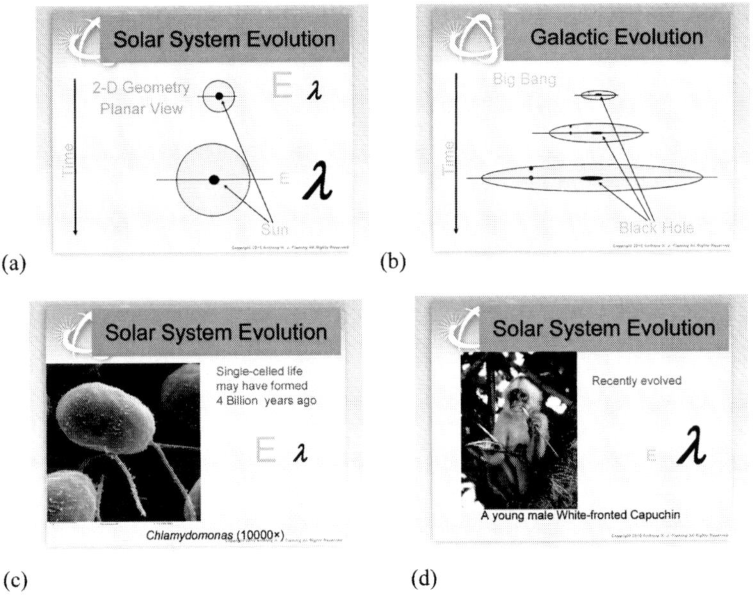

Figure 3.10 (a) Solar system growth in size and (b) galaxy growth in strength of structure as cosmological expansion proceeds. (c) Early life forms are small with a lower bulk modulus (invertebrates), (d) while recent species are larger with a higher bulk modulus (vertebrates). In the diagrams of galactic evolution the earth is shown as small coloured particles at two different times and locations, red near the GP and blue at its maximum excursion from the GP. As time evolves away from the big bang, the EM and A field resonances lose energy and increase wavelength; hence biological evolution proceeds towards bigger and more solid life forms with a drop-off in solidity near the points of maximum excursions.

the galaxy evolves cosmologically. This is generally what we see when examining the various phylogenetic trees of life. We would also expect to see a drop-off in solidification as the solar system oscillates up and down away from the GP, with an increase in solidification as the solar system moves near or at the GP.

Overall what is seen is the way rotations of the solar system are comparable with dendrochronology in that we can relate cycles of motion with events from other evidence. Hence dinosaurs become extinct at a point in time when the solar system was near a maximum excursion away from the GP. Dendrochronology is a very useful tool in reaching back into history and pre-history to fine-tune events within rotations of the earth around the sun that happened at the terrestrial domain. Cycles within the galaxy may also be linked to events within the fossil record.

About the dinosaurs and their extinction Wikipedia says, 'The Cretaceous–Tertiary extinction event, which occurred approximately 62.5 million years ago (Ma), was a large-scale mass extinction of animal and plant species in a geologically short period of time. Widely known as the K–T extinction event, it is associated with a geological signature known as the K–T boundary, usually a thin band of sedimentation found in various parts of the world.' It appears there was a specific extinction event and this was a major factor, but it may also be that in addition dinosaurs were structurally challenged by their size at that point of time in the galactic cycle, making extinction doubly certain.

The largest single living tree is ~1,800 tonnes, the largest living whale is the bull whale at ~190 tonnes, while the largest dinosaurs to have existed were less than half this. There is a gradation of size here that matches the pressure within the medium of the organism—soil, water and air. The ancestors of the dinosaurs include large flightless birds such as the ostrich, emu, cassowary and rhea. Did these birds lose flight as the pressure in the air diminished over the past galactic half-cycle, making flight impossible? Is there evidence of extinctions within the fossilised record of trees in the same way that dinosaurs become extinct around 62 million years ago? If not, why not? And why did whales survive to become the largest-known mammals and not become extinct at this same time? Why then were large trees and whales protected within the ground or the marine environment in a way that dinosaurs were not during an extinction event? It appears both replicate in a pressurised environment relative to air and this reduces the relative effect of any structural stress of the kind discussed in regard to dinosaurs, where life forms that existed totally in the air were not protected. Why did smaller terrestrial life forms survive? Why do astronauts get space sickness? In all these cases acoustic pressure within the galaxy appears to be a central factor.

Rather than survival of the fittest, a view at the individual scale of evolution, the view at the solar system, the galactic and other cosmological levels may involve resonance mechanisms and a synergy of fractal life forms evolving together. Life itself adapts to changing circumstances and is not the lonely survival for existence at the jungle level of existence. We may discover we are not alone; there may be life forms all around us in other solar systems and galaxies. Either we hide beneath our metaphorical blankets, afraid to look out beyond our own presumptions, or we go forward optimistically to forge our own destiny in the cosmos. The differences that exist between races, tribes or nations may be dwarfed by differences of species across solar systems and galaxies within the universe.

In the next chapter we examine various forms of diffusion, a process that appears quite chaotic and overwhelming in the milieu of the individual charged particles. It seems this picture, like Darwinian evolution, needs to be modified so that the underlying physics at the biophotonic level can be viewed. This reveals an SFT process at the cellular level that may be the basis of the cell cycle and the replication, growth and development of life forms. Thus life depends on cooperation across the very large domains within the cosmos and the very small domains of the cell and the (bio) photon.

Thus biological evolution appears related to cosmological effects:

(1) Biodiversity varied with the phonon energy of the galaxy, which expanded during the inflationary process. In general as expansion proceeded there was a steady increase in the bulk modulus of life forms over time.
(2) There was reduced viability of biological tissues to withstand the drop-off in bulk modulus at maximum excursions from the GP, leaving tissues less resistant to the forces wanting to implode their structure, especially large vertebrates. This reduction in bulk modulus may have left the largest of these species (dinosaurs) unable to function at a primary level.
(3) Biodiversity also depended on the energy density within the solar system as it evolved during inflation. Cells, tissues, organs, bone tissue and teeth evolved within the available energy density over time.

Thus as the solar system evolved it expanded and the energy the sun emitted dropped off over time and its associated wavelength also grew with time. Life forms evolved, as shown in Fig. 3.10.

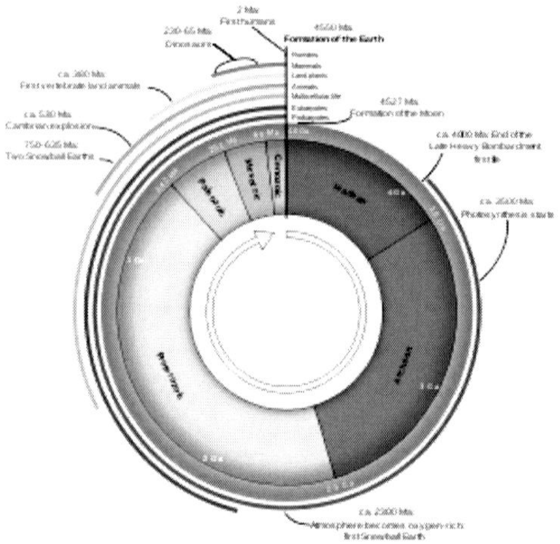

Figure 3.11 Geological clock, giving lengths of eons of the earth's history (credit: Wikimedia).

Biophotons that control mitosis and *biophonons* that control bone growth originate from far outside the body from celestial and galactic sources. This may form the basis of a spectroscopic test for extra-terrestrial life, given that the next-generation space-based optical telescopes may be capable of looking directly at exoplanets.

Shown in Fig. 3.11 a geological clock has been proposed where the clock face shows the eons across the earth's history. Given the cycles found within the marine species fossil record, the clock face might be given a period of 62 Ma. This matches the tri-spinorial basis for the solar system's motion around the galaxy. Again magnetic flips in the earth's geological record might also be used as a geological clock. If a quantitative basis can be determined for magnetic flips using SFT then an accurate record of past geological history could be generated to give a parallel dating to carbon-dating or dendrochronology, extending the historical range to either the earth's magnetic flip rate, mean rate ~100 ka (ka thousand years), or the galactic cycle of up and down wobbles of around 62 Ma. This would enable more accurate views of what are termed biological and geological time scales and cosmological evolution.

3.5 Cognition and Biological Evolution

Despite the most optimistic efforts of scientists, cognitive skills, including intelligence, memory and creativity, remain scientific mysteries and have resisted being squashed into theoretical boxes. Some artificial intelligence (AI) proponents have bravely suggested implanting lumped circuit memory chips into human brains as a way forward—that humans will soon become cyborgs, a presumptuous improvement on nature. Such fixes are at best simple solutions to complex problems. That said, AI is a valid means of making the simple, tedious tasks of life such as vacuum cleaning and driving easier and less onerous for humans. With a realisation of the immensity of the subject (Figs. 3.12 and 3.13), we carefully turn to questions surrounding cognition.

Why is intelligence such a recent evolutionary feature? Similarly, why are single-celled organisms and early forms of life apparently devoid of the intelligence we see in humans and primates? While cognitive skills, including problem solving, memory recall, social skills and language, can be seen in birds, mammals and marine species, the more cognitively able correspond to the more recent evolutionary genera, with hominids and humans at the top of the tree. From the previous discussion on the correspondence between the solar system and galactic evolution and biological evolution, we might ask what it is in the cosmological record that gives mankind this advantage. Cognition therefore appears linked to our biological structure. So what is different structurally between early forms of life and humans? At the next level of gravitation higher than galaxies, SFT indicates matter forms toroidal structures. Such cosmological structures do appear to exist as diffuse 'strings' of galactic matter. We note that the brain itself is a diffuse organ but not particularly in the shape of a toroid. However, this might not be the place to look for toroidal-shaped entities. Perhaps within the quarks there might be such entities; we simply do not know at this point in time. The purpose of the brain with its solid bone cranium is to securely hold and protect the soft tissues of the brain. The grey matter is the external structures of the actual boson-based information-bearing structures, including quarks containing the memory, and higher

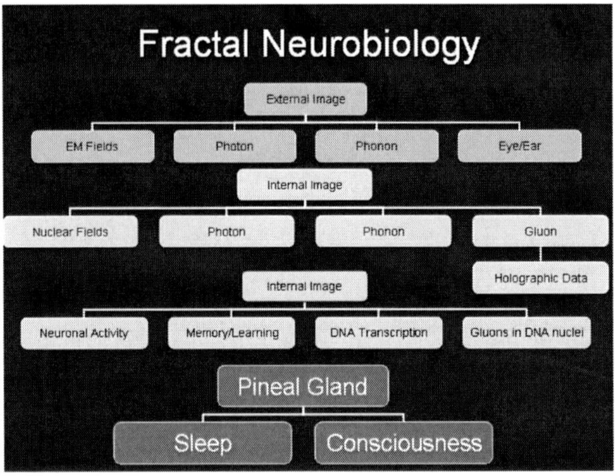

Figure 3.12 SFT concepts of consciousness being fractal and memories having field structures involving either biophotons (and/or) biophonons.

mental structures, which may reside in the DNA of the neurons (Fig. 3.14.)

There is strong evidence that our human memory has both short- and long-term components. SFT suggests that this may be linked to the way EM and A fields are stored within the (neuronal) DNA. It may be biophysically easier to store EM images as separate forms of memory than as collated images and sounds, which may require a separate biophysical pathway (Fig. 3.12). In brief, short-term memory may be linked to what we might call the EM atomic structure and long-term memory linked to the strong nuclear structure of the nucleus.

SFT sees life and consciousness as fractal processes having layered structures dependent on gravitation observed within the universe. Life on the earth evolved exposed to the gravitational fields of the moon and sun and the earth's vertical and horizontal components of E and H fields. Life also evolved due to the strong nuclear forces of our own galaxy. During our waking hours we are exposed mainly to the sun; during our nightly sleep, we are exposed to other astronomical objects, including the moon. Inert matter and various life forms are linked via boson streams to higher-level structures, including galactic clusters that may expose us to

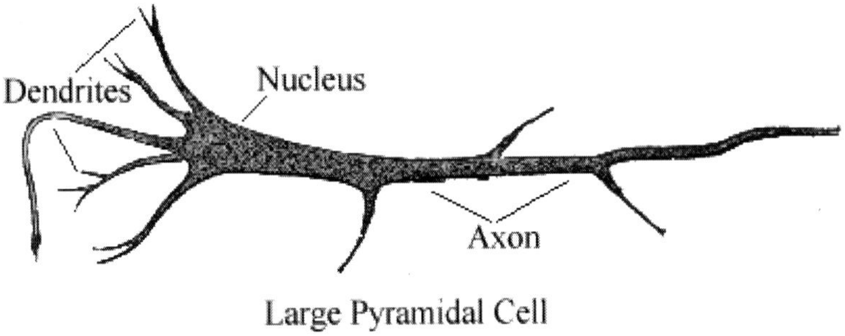

Figure 3.13 Diagram of the human neuron.

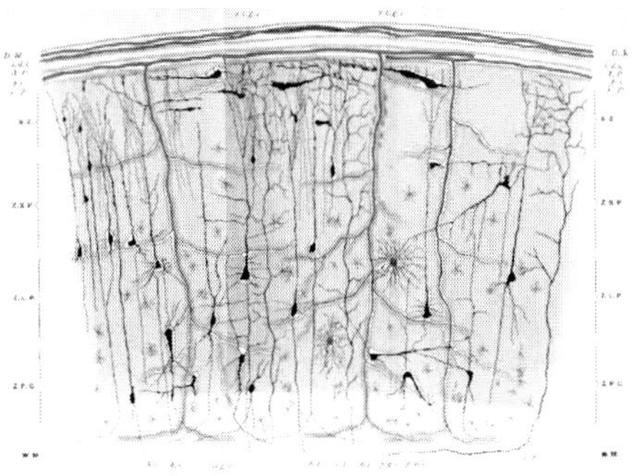

Figure 3.14 Structure of the cerebral cortex involved in memory, awareness, thought, language and consciousness, having six layers, each a different structure of neurons.

bosons related to these structures, and these may have their own evolutionary consequences.

Consciousness exists in different forms other than just our waking reality at the terrestrial level. Sleep research has revealed that the subconscious mind during sleep can carry out physical tasks that are not normal for humans. Many people dream of flying. One reported example is the reading of a clock held hidden on the ceiling above a sleeping person's body. This implies an ability to drift away from the body by a part of the consciousness, raising questions about the role of sleep. The description of sleep as a form of computer *defragmenter* gives only an electronic circuit analogy. Given the fractal basis of SFT's fields and particles, conscious existence may be layered; thought and consciousness like gravity may exist across a number of fractal field structures.

Rapid eye movement (REM) is known to be associated with the lucid dream state. This may be an indication that the body does indeed use the visual cortex within the brain during sleep and in particular during dreams. The eye muscles may still be operational within the visual experience during dreams. REM indicates a relativity in which the perception of elapsed time may be a neurologically internal reality.

Biophotons and biophonons may play a role in how mental processes are organised and how mental diseases come about. Communication between neurons may be compromised at times, for instance, with the onset of senility, especially dementia, when the biophysical structures of tissues are beginning to fail. Similarly mixed neural pathways may be an underlying biophotonic reason for drug-induced psychosis.

3.6 The Evolutionary Advantage of Dielectrics and Diamagnetics

When E fields in bulk tissues are measured using the methods of macrodosimetry, we find there are regions where the fields increase or decrease about their unperturbed free-space exposure levels and regions where the fields attenuate. These phenomena are related to polarisation and conduction of microscopic particles such as ions and proteins and structures such as membranes. As illustrated in Fig. 3.15 a dipolar molecule increases the E field at nearby points.

Thanks to the work of Maxwell, Wagner, Fricke and others, a macroscopic dielectric theory was developed so that Maxwell's macroscopic equations could be used and the microscopic details assumed without dealing with any underlying microstructure. While this form of classical EM theory remains useful for macroscopic calculations, including the determination of the specific absorption rate, it is incorrect at the atomic and photonic levels of interaction. Maxwell in 1881 calculated the equivalent conductivity of a suspension of homogeneous spheres (cells). Thus dipolar effects of similar microstructures even out over regions, and the concept

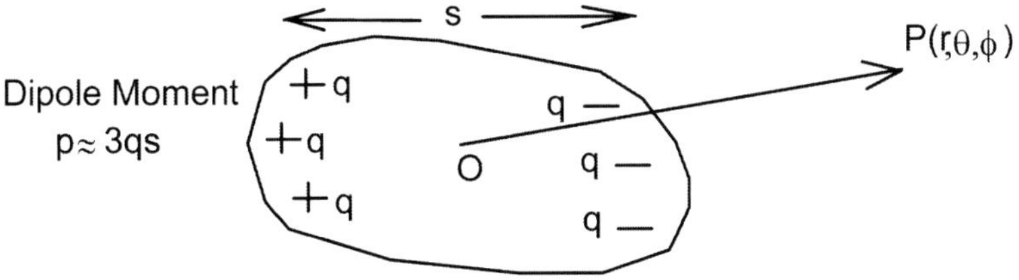

Figure 3.15 Small dipolar molecule. Approximate E field at point P in the far field of the dipole in spherical coordinates $E_r = \frac{2p\cos\theta}{r^3}$, $E_\theta = \frac{p\sin\theta}{r^3}$, $E_\phi = 0$ for $r \ll s$.

of permittivity is used to determine field effects in different types of tissues. For time-varying fields, complex forms are defined that incorporate conductivity. These bulk quantities are a measure of displacement and conduction currents, which are the motion or location of charged particles within the tissues. Similar comments apply to the measurement of B fields, and a volume-averaged, complex permeability can be used.

A macroscopic planar slab of homogeneous tissue is illustrated in Fig. 3.16. Foster and Schwan in 1986 showed how dielectric quantities can be measured by circuit analysis concepts and modelled as lumped circuit elements of capacitance and resistance. Actual measurements are plagued by leakage fields, and hence there are inaccuracies in such measurements. After it has been charged, any lossy tissue exhibits a relaxation (or dispersion) of charge, which is observed as an exponential decay with time. Foster and Schwan described general relaxation theory, whereby in most

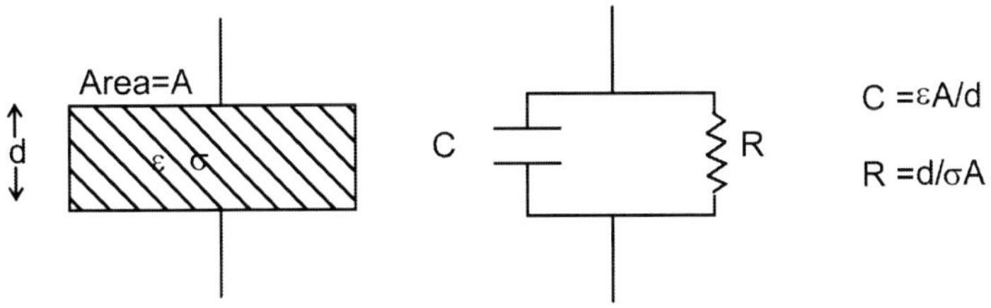

Figure 3.16 Planar homogeneous slab of dielectric tissue; dielectric properties measured macroscopically via attached electrodes and modelled as lumped circuit elements.

real tissues there is a distribution of these dielectric dispersions across the spectrum, which can be represented mathematically by a superposition of dispersions.

$$\varepsilon(\omega) = \varepsilon_\infty + \Delta\varepsilon_1 [1 - \exp(-t/\tau_1)] + \Delta\varepsilon_2 [1 - \exp(-t/\tau_2)] + \ldots \quad (3.8)$$

According to SFT, measurements of a planar slab, as illustrated in Fig. 3.16, reveal both conduction and magnetic processes in biological tissue. Both are forms of imaginary impedance; one is positive impedance, while the other is negative. Hence if both processes are present, then the measured result will be overall impedance, the difference between the two.

The various dispersions, both permittivity and conductivity associated with a particular tissue, are the result of differing underlying biophysical mechanisms. It is common that there exist three or more dispersions termed alpha, beta, gamma and delta (α, β, γ, δ) in tissues. For instance, relative permittivity decreases with increasing frequency in three or more major steps. α-dispersion ($\sim 10^2$ Hz) is associated with polarisation within membrane surfaces. β-dispersion ($\sim 10^6$ Hz) occurs due to capacitive charging of cell membranes in tissues, with a small component due to protein dipolar relaxation. γ-dispersion ($\sim 10^{10}$ Hz) is due to the dipolar relaxation of water molecules. Study of underlying mechanisms for dielectric dispersion is an active area of research (Fig. 3.17).

The clinical uses of measuring the intrinsic electrical and magnetic properties of biological tissues include impedance tomography, X-ray tomography and nuclear magnetic resonance imaging. What surprises in AD 2011 is the comparison between dielectric and diamagnetic measurements and theory. A general assumption within the bioelectromagnetics (BEM) community is that for many biological tissues there are no differences to the diamagnetic measurement of free space. This assumption is certainly incorrect within the DNA with its various structural forms. Yet BEM clinicians and researchers have largely ignored this line of research to date. There appears to be a wealth of practical uses for magnetic fields in clinical situations, including bone refracture, perfusion and prophylactics against stroke, amongst many.

While these macroscopic measurements and dielectric theory have discovered much about biological dielectrics, the underlying

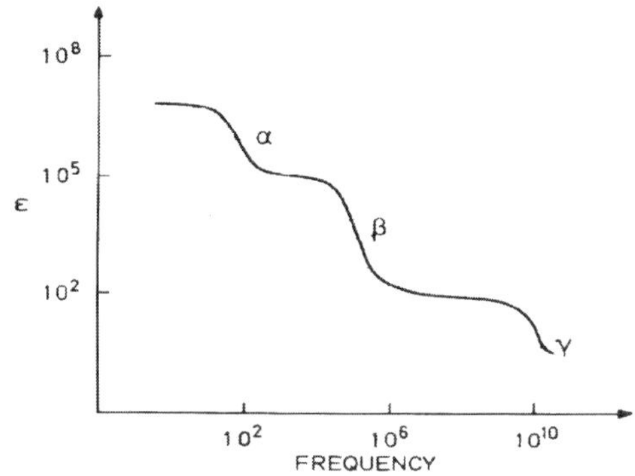

Figure 3.17 Dispersion in biological tissues as α, β, γ regions (*source*: Foster K.R., and H.P. Schwan, Dielectric properties of tissue, in Polk C., and E. Postow (Eds.), *Handbook of Biological Effects of Electromagnetic Fields*, CRC Press, Florida, 1986).

photonic reason for their existence has not been considered. Fleming has suggested that the physical cause of α-dispersion is due to the rotation and alignment of cells within tissues that can magnify E fields inside tissues and in the cytoplasm of cells. At static fields this is a cooperative mechanism by which cells act in concert to help achieve sufficient quanta of energy in order for fertilised cells to continue to achieve complete mitosis. This is a mechanism by which biological evolution has found to achieve mitosis. This cooperation between cells is almost the opposite of the 'survival of the fittest' mechanism that Darwin found underlying the origin of species. Perhaps this is result of cycles within evolution. There appears to be a similar diamagnetic mechanism by which enhancement of magnetic fields can be used by the loops of DNA to help achieve mitosis within cells. We discuss this combined E and H field mechanism in Chapter 5. This combination of E and H fields in cellular dynamics is similar to the combined E and H fields used by electrons and protons within atoms to achieve stability as seen by SFT. The reason for biological dielectrics and diamagnetics is as a means of achieving mitosis to counter the effects of evolutionary expansion. This has been an evolutionary advantageous method whereby replication of species has come about and was apparently in place from the time single-celled organisms came into being, from the very earliest stages of life on earth.

The universe and its evolution since the big bang can be thought of at the EM level as a varying dielectric and diamagnetic object of immense scale. As biological evolution on the earth began, the energy density within the universe dropped considerably. We see the need for life to evolve cooperatively. Cells not only behave cooperatively, but they have cellular tissues around them to maintain the energy densities they required in past stages of evolution. This is an underlying reason, namely, to create higher-energy densities via cell–cell cooperation. What can also be seen is that not only dielectric magnification but also diamagnetic magnification can be used as an evolutionary advantage since both electric and magnetic fields have diminished due to the universal expansion since biological evolution began. Indeed some lower organisms, including bacteria, show a pronounced sensitivity to the earth's magnetic fields. This also appears to be the case in cellular dynamics discussed in Chapter 5.

As recently as 1987 the World Health Organization (WHO) in 'Health Criteria Document 69: Magnetic Fields' published a comprehensive review of the effects of static and time-varying H fields, finding the following conclusions:

(1) Only a few mechanisms of the interaction of biological tissue with H fields have been established. Some of the biological effects data suggests that other mechanisms may play a role, but these have yet to be confirmed experimentally. Thus, only a preliminary assessment of the human health risks from exposure to H fields can be made.
(2) A number of lower organisms have shown a remarkable sensitivity to the earth's H field, because of highly developed receptors. Similar receptors have not been found in human beings.
(3) For human exposure to static H fields, it is not possible to make any definitive statement about the safety or hazard associated with short- or long-term exposure to fields above 2 T. Available knowledge suggests the absence of any measurable effect of static fields on many major developmental, behavioural or

physiological parameters in higher organisms. Recent medium-term (days) studies on exposure of animals to static fields of up to 2 T have not demonstrated any detrimental effects.

(4) From the scientific database on higher organisms exposed to H fields, only four types of effects can be regarded as established. The first three may be explained by plausible mechanisms of interaction and produce a basis for extrapolation to man. These effects are:

(a) induction of electrical potentials and magnetohydrodynamic effects within the circulatory system;
(b) the formation of magnetophosphenes with a time rate of change of magnetic field exceeding 0.3 T/s at 17 Hz; the effect depends strongly on frequency;
(c) direct stimulation of nerve and muscle cells by very short (less than 1 ms) pulses of rapidly changing magnetic fields (several thousand T/s). Current densities are estimated to exceed 1,000 mA/m^2. These effects are strongly frequency-dependent and may exhibit lower thresholds (100–1,000 mA/m^2) under more favourable stimulus conditions (10–100 Hz); and
(d) other cellular and tissue alterations when the induced current densities exceed approximately 10 mA/m^2 (http://www.inchem.org/documents/ehc/ehc/ehc69.htm).

There is a constant theme running through classical EM, quantum theory and the research and clinical uses of BEM. As Einstein insisted about quantum theory, they are all incomplete; there was a lack of magnetic theory and measurements throughout the 19th and 20th centuries. With no magnetic theory to use at the photonic level, the measurements have had no direction and no known mechanisms to look for in regard to magnetic fields. On the other hand, SFT, its theory of evolution, both cosmological and biological, and its theory of the atom all provide missing magnetic fields. This is the case with the Bohr theory of the atom, for instance, where SFT provides the missing H fields, forces and motions. The atom can now be balanced without recourse to the incomplete quantum theory. The BEM community, like the EM community, faces the reality it is based on an incomplete classical EM theory. Given the recent appearance of SFT within science, the BEM community will need to embrace SFT with its complete EM theory. Then H field effects may eventually be seen as potentially harmful if they induce rotation in the same way that E fields are seen to move particles due to translation. In Chapter 1 we examined a case where the cell nucleus was seen to rotate due to exposure to a radiofrequency field. In this case cancer cells were being rotated. If on the other hand normal, healthy cells and their nuclei were similarly being rotated then this could be medically injurious to the health of the person, especially where long-term exposures are involved. Tissues that evolved in environmental levels of EM exposure need to be protected from overexposures. In Chapter 4 we examine how exposure to H fields can cause rotations of charged or dipolar particles.

3.7 Global Warming and Cooling Due to Galactic Tides

In this chapter we have been looking at how biological structure and evolution relate to gravitational structure and cosmological expansion. There is another very important phenomenon that relates specifically to galactic motions, namely, global warming. SFT sees a series of energies that oscillates between negative and positive across the universe, depending on the analytic form of gravity acting within that part of the universe. The sun is a positive source of energy, while the galaxy's black hole is a source of negative energy. The solar system bobs up and down in its motion around the galaxy (see Figs. 3.5 and 3.10), and this warms the earth at its maximum excursions from, and cools near, the GP, a cycle of around 62 Ma. This natural warming and cooling cycle due to the negative energy of the galaxy cannot be separated from global warming due to atmospheric carbon dioxide by overuse of fossil fuels such as coal, petrol, propane and methane. This has been the case since the start of the Industrial Revolution in the late 18th century and presents a significant man-induced risk of a runaway greenhouse effect. Mars is thought to have once suffered a wholly natural greenhouse runaway effect, leaving it by and large a sterile planet; SFT sees this more as a result of an expanding universe. While SFT provides evidence the earth's warming is part of a natural cycle it would be imprudent in the extreme to ignore the risks and induce a catastrophic avalanche into an avoidable and perhaps irreversible change of climate. The atmosphere and oceans of the earth are not inexhaustible resources. There are signs that cities can become technological 'hot spots' due to the centralisation of energy sources. Perhaps by using the closed-form solutions given by SFT, nuclear fusion energy might be produced in a safe manner and engineered to be small enough to provide power at the small town or even single-house level, enabling a decentralisation of energy. If this is the case, then the emitted H, E and A fields could be shielded within associated metal and μ-metal housing and cabling, and acoustic baffling, so humans are protected from spurious exposures above environmental levels. Mankind needs to generate energy via a method that is free of carbon dioxide and safe. Perhaps the mathematics of SFT can reveal some such nuclear method.

Chapter 4

Biophotons and Diffusion in Biology

Intimately involved with growth and development biological diffusion includes random motions and non-random electric (E) and magnetic (H) field offsets of charged and dipolar entities. Diffusion applies to biology generally, to the cell cycle and the body's mechanisms for repair and maintenance in particular. Diffusive phenomena concern particles of relatively large atomic mass and biophotons of tiny mass. Self-field theory (SFT) brings fundamental factors to the theory of diffusion. Having internal structure biophotons can change state abruptly, inducing cascades at discrete energy levels and frequencies. In SFT E and H fields are opposite forms of motion, analogous to charge. H fields remove energy, while E fields add energy, for example, the earth's polar zones are regions of high H fields, where the earth loses energy, while the equatorial zone is a high-E field region, where the earth gains energy. Chemical cascades occur at 'thermal' energies, for example, rain formation via storm clouds, and 'non-thermal' energies, for example, cell division. Mechanisms of growth and repair include cascades. Various effects are excluded from classical electromagnetics (CEM) heat analyses. But Planck's blackbody theory applies to all frequencies. In many viscous models of diffusion motion is assumed translational, ignoring rotational effects. Theory and observations of rotation of

Inside the Photon: A Journey to Health
Tony Fleming

ISBN 978-981-4241-40-3 (Hardcover), 978-981-4241-88-5 (eBook)
www.panstanford.com

cell nuclei or proteins within membranes help explain window observations of Adey at low levels of extremely-low-frequency (ELF) exposure, significantly below international standards of exposure limits. The physical mechanism is the EM excitation of a dipole, similar to SFT's dipole–dipole gravitational interaction. Both physical mechanism and its biophysical analogue support Adey's window effects in embryonic colonies.

4.1 Introduction

Diffusion of charged and dipolar particles is intimately involved in a wide range of biological processes. Ions and proteins amongst other biological components have long been known to diffuse. In many cases, present knowledge is incomplete, yet enough is known to describe the basic mechanisms behind biogenic division, which we investigate separately in Chapter 5. Figure 4.1 depicts four examples of effects due to static electric (E) and magnetic (H) fields where ion or protein diffusion (or diffusion of both) is involved. Figure 4.1a shows a shark electrosensitive to weak E fields, 0.5–1 $\mu V\ m^{-1}$, emitted as ion currents by prey and magnetosensitive to the local geomagnetic field (LGF) via larger E fields, 5–50 $\mu V\ m^{-1}$, induced by the flow of ions in ocean currents. Figure 4.1b shows a *spirillus* alga, a single-celled creature that moves north–south along the LGF, possessing its own biogenic magnetite crystals. It is known there is an induced torque on the magnetosomes, membranous structures containing the crystals imbedded in the creature. Torque is induced by the H field near the crystals and the algal motion. What role ionic currents may play in magnetotaxis is unknown. Self-field theory's (SFT) photonic theory is consistent with the earth's magnetic flips and the direction (north and south) of these magnetically sensitive motions. These flips of the sun and earth are a form of the cascades seen in chemical and biochemical reactions. These flips are a form of photonic structural change that occurs as the photon changes its energy state.

Figure 4.1c shows a membrane interfacing a cell surface. Ionic concentrations inside and external to the cell are controlled by the permeability of the membrane to various ion species

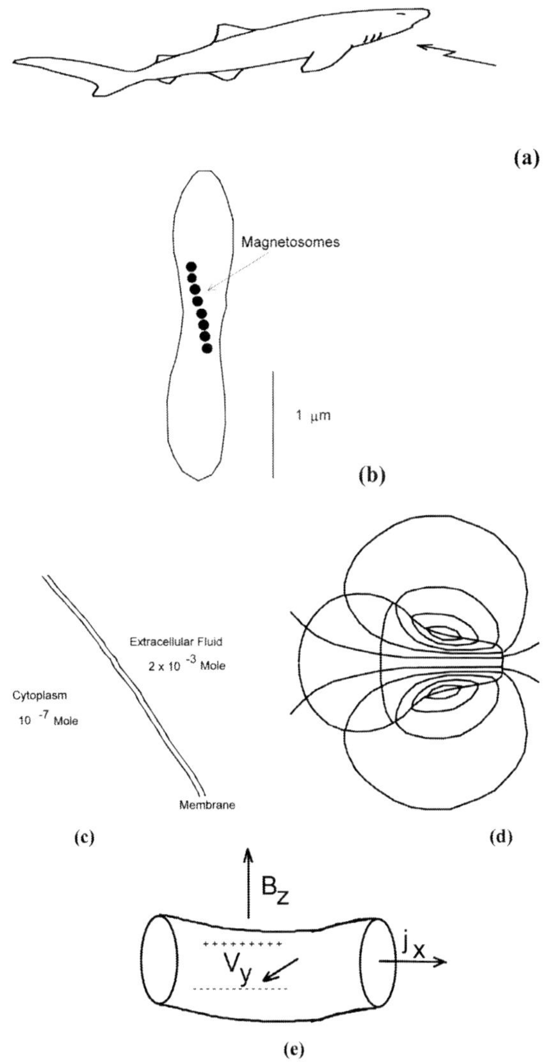

Figure 4.1 (a) Shark sensing diffusion of ions emitted by wriggling prey. (b) *Spirillus* alga using torque induced by the LGF to navigate in the direction of the LGF. The record of such motions is revealing the palaeomagnetic history of the earth. (c) The membrane across the biological interface allowing control of ion diffusion. (d) Cell formation of the seaweed *Fucus* controlled by ion diffusion in and around the cell. (e) Induced potential due to ionic motion used to measure blood flow in aortic vessel.

Shown are typical calcium ion concentrations measured inside and outside cells. As a final example, Fig. 4.1d depicts ionic currents experimentally measured by Jaffe, who devised an instrument known as the vibrating probe and detected current densities, 10–100 μA m^{-2} in extracellular regions and near the growth tip of the seaweed *Fucus*. Such measurements reveal the presence of biogenic fields that must be involved in generating these currents. Similarly diffusion of proteins within membranes is considered due to biogenic fields. In turn such diffusive processes are considered to be involved in cell growth and development. Many experiments have studied such effects using techniques to track the in situ migration of membrane components. Figure 4.1e shows the transverse potential difference, V_y, that exists in a capillary vessel due to the flow of charged particles within it across a magnetic field.

The diffusion of charged particles such as ions exposed to E and B-[a] fields forms a large body of research (Fessard, 1974; Polk and Postow, 1986; Nuccitelli, 1986; World Health Organization [WHO], 1987; Carpenter and Ayrapetyan, 1994). Diffusion due to moving media also occurs in biology. Diffusion of charged particles other than ions occurs in biology. Poo discussed the mobility of dipolar proteins in plasma membranes due to applied E fields (1981). A freeze fracture of a vesicle of the mitochondrial inner membrane clearly shows that the proteins diffuse in the direction of an applied voltage gradient (Sowers and Hackenbrock, 1981). Measurements show that the rate of this diffusion is similar to that of many other proteins in a fluid mosaic membrane. The lipids that comprise a large component of the plasma membrane and other membranes also rotate and diffuse (Darnell et al., 1986).

Some bioeffects are dependent upon the microstructure and need to be studied using microdosimetric techniques, as discussed by Tenforde. It is no surprise that there are effects due to the biological microstructure that are not measurable using macroscopic methods. Figure 4.2 shows a schematic of a human cell. New methods of electron microscopy are being developed and used to

[a]We sometimes use the abbreviation 'B field' to specifically distinguish from the H field. While most biological tissues have a permeability μ reported to be no different to that of free space (μ_0), some such as the histones of deoxyribonucleic acid (DNA) have an intrinsic magnetic structure.

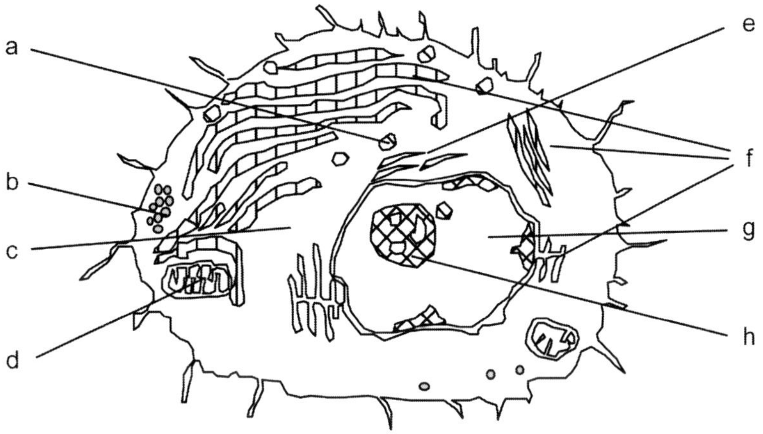

Figure 4.2 Schematic showing human cell microstructure: (a) lysosome, (b) glycogen, (c) ribosomes, (d) mitochondrion, (e) Golgi apparatus, (f) endoplasmic reticulum, (g) nucleus and (h) nucleolus.

uncover previously unrevealed cell structure of exquisite complexity. Penman discussed how these new images are demanding a rethink of our previous ideas of cytoskeleton, nuclear matrix, mitosis and the relation of membranes to cytostructure. Grynkiewicz et al. and Wolke et al. have used fluorescent markers to determine cellular concentrations of ions. Sowers and Hackenbrock and White et al. have used these same methods to study the motions of cell components. In line with these advances, concepts of biological and cellular interactions with E and B fields based on simple geometries, for example, layered spheres and ellipsoids, also need reworking.

In a 1994 review of E and B field bioeffects, Carpenter and Ayrapetyan detailed the need for research into the influence upon biological cells of both static and time-varying fields. Matters of concern and opportunity in the areas of biohazards, biomedicine and other bioelectromagnetic (BEM) and biophotonic areas, such as developmental and plant cell biology, need resolution. Glimpses of the possibilities exist, such as the use of pulsed electromagnetic fields (PEMFs) and pulsed magnetic fields (PMFs) in bone refracture therapy and high-intensity PMFs as a prophylactic against the often fatal damage caused by ischaemic attacks and to mitigate against pain.

Over the last several decades, the body of knowledge of the influence of endogeneous and exogeneous E fields upon cells has been expanding rapidly. A range of weak, medium and strong effects is now partially understood. The range of effects observed on single cells due to E fields includes electrophoresis, the translation of whole cells in the direction of the applied field; dielectrophoresus, the movement of cells in a dielectric medium associated with time-varying fields; electrorotation, the rotation (alignment) of cells in polarised fields; membrane electrophoresis, the diffusion of components within the plasma membrane; and electrode formation, a distortion of the shape of the cell due to membrane stresses.

In many of these effects, a quantitative understanding has emerged, and theory can be tested against computation and observation. However, no complete picture has emerged of the way in which E fields operate to affect cell function. While 'snap-shots' of theories exist, numerical methods based on a three-dimensional cell structure might provide a more meaningful 'moving picture' of these effects. A model that can calculate in the primary unit of cell effect, namely, any changes in charged particle flux flowing in and around cells, and then obtain any flux-dependent effects, would be most instructive. As distinct from one-dimensional (circuit analysis–type) calculations, which offer only an 'on-off' answer, there may be subtle effects revealed by detailed three-dimensional numerical analysis of a process that evolves over time. Simple three-dimensional models of cells of this kind have begun to emerge. Sala and Hernández-Cruz (1990) examined the intracellular calcium inside a spherical model of a neuron due to buffers and tracers. No complete model that includes the membrane is available—certainly nothing to assist in the study of E and H field effects.

There are several interrelated research areas concerned with electrical exposure of cells. In the following, we survey three reported non-destructive bioeffects with an eye towards the eventual implementation of a computational approach towards cell biology. There is a natural progression through the three, from direct electrical contact, through medium to strong non-contact E field exposure, to weak E field and ultraweak exposures.

4.2 Direct Contact Effects: Membrane Voltage Patch Clamp

Some cells act together to form physiological systems such as the heart, nerve fibres, muscles and neuromuscular junctions. When the membranes of these cells are electrically fired, ions are released across the membrane. This occurs when a certain membrane potential has been reached. Commonly the cells are then able to propagate an ionic signal to their neighbours, which are physically adjacent, and a macroscopic effect follows. Electrophysiology is the study of such cellular systems that oscillate between resting and active states and are controlled by voltages applied across their membranes.

The study of membrane biophysics owes much to the pioneering work of Hodgkin and Huxley, beginning in the early 1950s (Plonsey and Barr, 1988). Using voltage patch clamp electrodes, they investigated membrane ionic currents and their dependence on transmembrane potential. Starting with a resting membrane, in dynamic equilibrium, they increased the transmembrane potential until the cell became active, deriving equations relating the time-varying ionic currents and the membrane potentials. Their work has led to a deepening quantitative understanding of transmembrane dynamics (Chay and Keizer, 1983; Chay and Rinzel, 1985). The stochastic gating mechanism of individual ion channels is now understood so that this near molecular analysis ties in with effects at the cellular level (Plonsey and Barr, 1988). This direct contact effect has in turn led to more recent experiments trying to elucidate the corresponding transmembrane ion dynamics when cells are exposed to external E fields (Ehrenberg et al., 1987).

4.3 Biogenic E Field Effects[a]

The growth and development of cells are thought to be partly controlled by bioelectric fields. While the strong effects of electro-

[a] There is a rich history of research into bioelectric fields and cells, going back at least to the discovery of the cell as a fundamental unit of biology by Schwann and Schleiden in 1847 (Lund, 1947). It is worth noting that since Mesmer in the

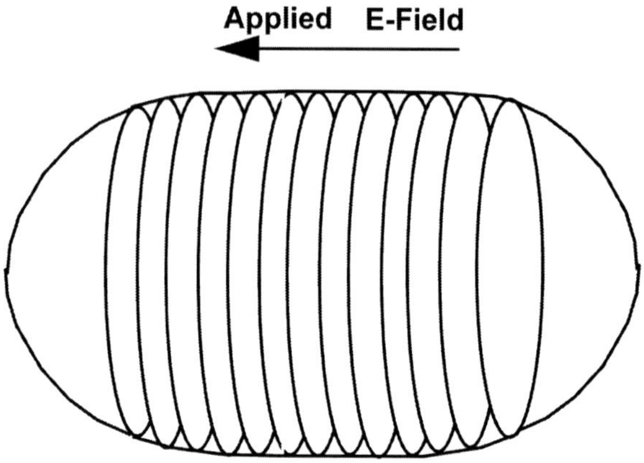

Figure 4.3 Growth in plant cells. Left–right symmetry of regrowth of the *Mougeotia* protoplast wall.

physiology are theoretically and experimentally indisputable, the weaker effects of biogenic E fields are yet to be fully understood. The effects discussed below may well partially depend upon the flow of ionic currents near cells (Jaffe, 1979) and upon protein diffusion within plasma membranes (Poo, 1981).

In a recent study, spherical *Mougeotia* protoplasts, originally cylindrical, were exposed to E fields of 20 V m^{-1} (White et al., 1990). Over some hours, the protoplast walls regenerated, re-establishing the original shape. How regrowth occurs is unclear; but one possibility concerns electrophoresis of membrane proteins (Poo, 1981) (Fig. 4.3).

Effects of exposure to static E fields of 1,000 V m^{-1} upon the cytoskeleton were noted in experiments designed to observe and quantify galvanotaxis of fibroblasts (Hui, 1994). Changes in intracellular calcium flux were measured using aequorin, a calcium chemiluminescent indicator protein, and fluo-3, a calcium-sensitive fluorescent dye. Motility of the fibroblasts was measured,

1770s until our own time, those who attempted to claim biological effects due to magnetism, 'animal magnetism', were often denounced as quacks and frauds, in total contrast with the popular success of experiments by Galvanni and Volta to demonstrate 'animal electricity' (Frankel, 1986). It appears an understanding of biological effects due to magnetic fields has at last come in from the cold. Biomagnetic effects are often more subtle than bioelectric effects.

fluorescent micrographs showing changes in actin-containing stress fibre patterns of the cytoskeleton that span across the cell body and are attached via proteins to the inner plasma membrane.

4.4 Weak and Windowed EM Exposure Effects

on cell division

Over the past decades, a number of reports have emerged of weak, extremely-low-frequency (ELF) and modulated ELF radiofrequency (RF) exposures inducing measurable ionic changes in cells. Windowed effects at power densities less than 1 mW cm^{-2} across both ELFs and RFs have been reported (Adey, 1980; Adey and Sheppard, 1987; Bawin et al., 1989). In the earliest experiments, chick brain hemispheres in saline solution inside a test tube of about 1.4 cm diameter were exposed to ELF and modulated ELF RF fields in air (Fig. 4.4a,b). Extracellular tissue fields at ELF are estimated to be 10^{-6} V m^{-1} (Fig. 4.5), and the corresponding RF levels are around 10 V m^{-1}.

Adey's window experiments resulted in frequency-dependent changes in intra- and extracellular Ca^{2+} (efflux/influx) concentration levels both at ELF and modulated ELF frequencies (Bawin et al., 1975; Bawin and Adey, 1976; Liboff et al., 1987). The original observations in which in vitro specimens were exposed in air remain controversial. Since the observations were not understood mechanistically using classical electromagnetics (CEM) theory, there was question within the bioeffects community about the validity of these effects at weak exposure levels. There was further controversy as the modulated effects appear at levels below those set down by national and international RF exposure standards, such as the International Radiation Protection Association (IRPA), the Standards Association of Australia (SAA) and the American National Standards Institute (ANSI), while ELF effects occur at much weaker levels. SFT in contrast to CEM suggests that such effects may be mechanistically understood and suggests further that the effect at these weak levels may indeed constitute a possible health hazard and is not just an interesting bioeffect. This opinion is related to possible effects upon cellular division. Public concern demands further study to provide a clear and quantitative understanding. In Chapter 6 we examine the use of PEMF-based therapies that

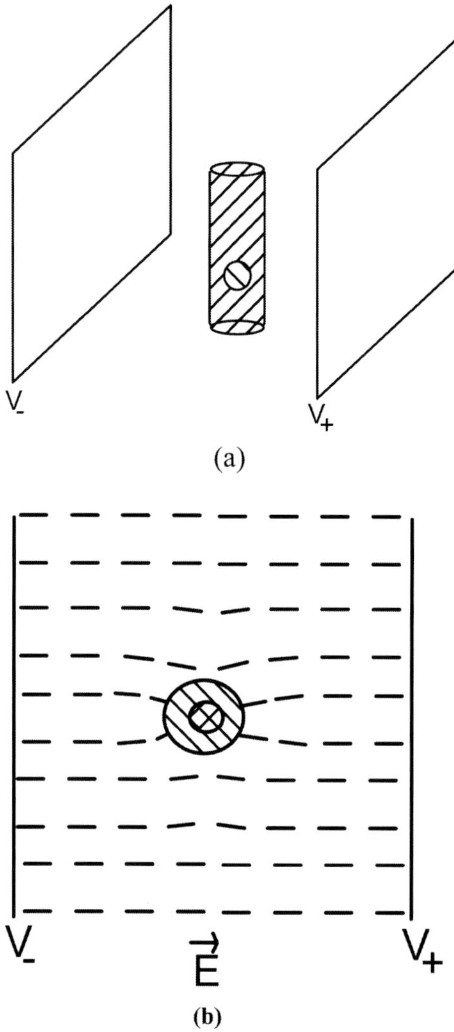

Figure 4.4 (a) Schematic of in vitro ELF exposure set-up ($E \approx 50$ V m^{-1}). (b) Plan view of exposure set-up showing lines of potentials acting on tissue between plates.

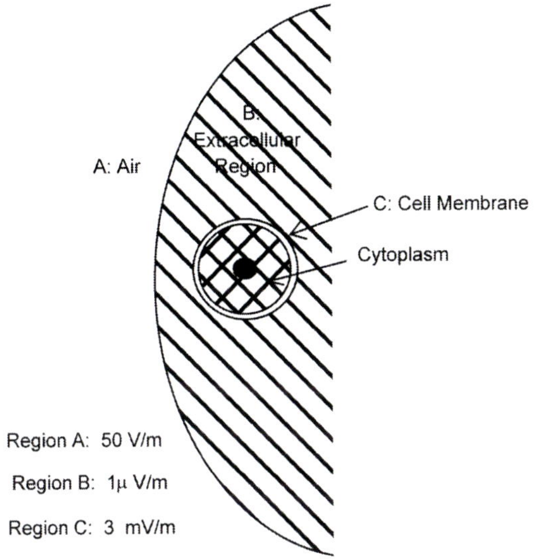

Figure 4.5 Fields in an extracellular medium and cell membrane estimated due to 50 V m^{-1} E field in air where the body's surface is treated as having ubiquitous resistivity (no significant biophoton flow across the epidermis or its pores).

operate at such weak levels, well below those recommended by international and national exposure standards. If such levels can produce medically proven health benefits it is probable they may also produce harmful effects. Further research is needed.

What some in the bioeffects community have not previously accepted is how membrane potentials of around 100 mV relate to such experiments, since to them this implies exposure levels in air above air's dielectric breakdown strength (Adair, 1991a). The use of CEM along with simplistic tissue geometries lies behind such computational overestimates. While CEM sees circuits as the primary focus of attention, biophotonics, as we have discussed, is a lower level of interaction between atoms, ions, proteins and macromolecules. There are areas of the anatomy particularly relevant to biophotonic flows. Pores of conductivity in the epidermis are relevant in ELF and modulated ELF experiments.

Pores within the epidermis are a basis for acupuncture and allow biophotons to enter or leave the body. Acupuncture, like

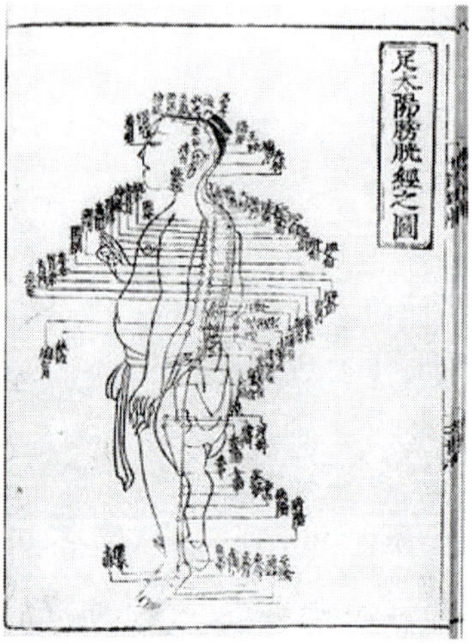

Figure 4.6 Acupuncture chart 1716 (credit: Wikimedia). Acupuncture is based on the flow of biophotons across the skin.

homoeopathy, is an alternative therapy whose biophotonic basis appears important to its efficacy (Fig. 4.6). Such traditional therapies are assumed to act as placebos by some scientists. SFT allows fresh perspectives linked to mechanisms available to the biophoton. The concept of tensegrity, fibres running throughout the body, is a relatively little known but important area of medical research. Microtubule fibres are an element of the overall anatomy of species used for structural integrity, dynamic movement and signal pathways within an organism. There are links between hair growth, pores of the skin and microtubules. Microtubules, their energy state and flexibility may play an important role in conditions of old age. Deterioration leading to loss of energy may lead to loss of function of these fibres. Knowledge of photon states and binding mechanisms allows an understanding of how biophotons can cause macroscopic chemical changes such as cascades in ion concentration levels.

SFT was recently used to examine possible internal structures of the photon. One possible structure of the ordinary stationary photon is hydrogenic, as shown in Fig. 4.7; two sub-photonic particles termed the 'ephectron' and the 'phroton' have equal mass and opposite charge. This provides Einstein's photoelectric energy

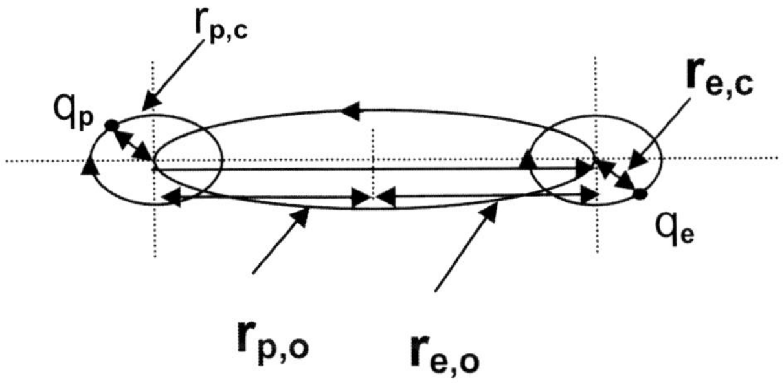

Figure 4.7 Composite photon. SFT provides a rationale for the differences between (a) quantum physics having a discrete energy spectrum as its sub-particle masses are unequal; only certain radii and velocities are valid solutions to the SFT equations; and (b) continuous physics having a continuous energy spectrum as its sub-particle masses are equal; all radii and associated velocities are valid solutions to the SFT equations.

$E = h\nu$ and gives the proper continuous energy–frequency response, unlike the discrete physics of the atom where the electron and proton have differing masses. The sub-photonic E and H fields for the two interacting particles are calculated using a SFT formulation similar to that applied to the hydrogen atom. SFT applies to the photon in the same way it applies to the hydrogen atom. The primary parameter in the analytic spectroscopy of the ordinary photon is its mass, similar to the principal mode of the hydrogen atom where the electron mass specifies the spectroscopy. Where the photon mass is known, the spectroscopy of photons can be examined in detail. The transition frequencies are expressed in terms of a continuous series,

$$\nu_{mn} = R_{\text{ph}} \left(\frac{1}{m^2} - \frac{1}{n^2} \right), \tag{4.1}$$

where $m = 1, 2, 3, \ldots; n = 2, 3, 4, \ldots$

The photon Rydberg number

$$R_{\text{ph}} = \frac{q_{\text{ph}}^4 m_{\text{ph}}}{8\varepsilon_0^2 h^3 c} = \frac{m_{\text{ph}}(8^2\pi^3)\varepsilon_0 v_{\text{o}}^2}{r_{\text{o}} q_{\text{ph}}^2} \tag{4.2}$$

is a photon-specific function of mass and charge. Such stationary photons may probably be found in the outer extremities of the universe. In this case, the photon has four degrees of freedom. In the terrestrial domain, photons move as the binding energy of atoms and molecules at the speed of light where they now only have two independent degrees of freedom; these form two extra quantum numbers.

One fundamental mechanism of this hydrogenic photon structure is a sudden dramatic change of state as energy transition levels are transversed. This is related to cascades observed in reaction rates associated with chemical kinetics of many physical and biological processes. These cascades form an important role in the body's methods of healing and repair, for instance, heat shock protein. Intracellular heat shock proteins are being trialed for the treatment of several types of cancer, indicating a beneficial therapeutic role for ultra-weak exposures.

4.5 Magneto- and Electrosensitivities in Birds and Marine Species

Biological life forms, humans, birds, fish, insects, bacteria and other systems, exist within an EM milieu, reacting at the cellular level to endogenous EM signals. As to exogenous exposures, humans are not overtly sensitive to environmental E or B field levels, unlike many birds and marine species that have evolved physiological methods to use biogenic and environmental fields for everyday purposes. In the early proto-earth the atmosphere was warmer and more hydrated than it is today; mammals, including humans, are thought to have evolved from marine forms of life where the EM environment was of larger magnitude because of the surrounding conductivity of the marine milieu. Certain flying insects and birds are observably magnetoreceptive, navigating and orienting themselves according to the LGF (Keeton et al., 1974; Adey, 1981). In the absence of other cues, bees orient their hives according to the LGF, showing acute sensitivity to temporal change as small as 1.5×10^{-10} T min^{-1} in the LGF. Migratory and homing abilities of birds may depend on multiple cues, including visual landmarks, the position of the sun and other cues; statistical evidence suggests a discriminatory behaviour with respect to the LGF.

Magnetoreception is also observed in the marine and amphibian environments. Certain single-celled bacteria and algae exhibit motion parallel with the LGF and possess chains of magnetite crystals oriented linearly with respect to their cylindrical outer membranes (Kalmijn, 1981; Frankel, 1986a, 1986b). Sharks when swimming across the LGF sense the potential difference thus generated (Kalmijn, 1981). Rays have been shown to use the LGF to determine the direction of their circular path around a tank aligned symmetrically with the LGF. During migration, whales exhibit magnetosensitivity, and temporal changes in the LGF due to solar activity and aperiodic changes in the earth's mantle may cause the well-known beach strandings of schools (Walker et al., 1986).

Befitting the conductivity of the marine environment, electroreception is also observed, the local electric field (LEF) being used for sensing purposes (Kalmijn, 1974, 1981; Adey, 1981). Marine vertebrates such as sharks and rays exhibit predatory feeding behaviour that depends upon some LEF emitted by the prey. Flatfish may attempt to escape by burying themselves in the sand on the seabed, but this does not sufficiently camouflage their bioelectric fields to avoid well-aimed attacks by both sharks and rays (Adey, 1981). Indeed, the wriggling motion of flatfish requires the firing of electric signals in its muscles, which create the fields that are then detected by the predator.

Turning to biogenic methods of field amplification, there is evidence for LGF amplification amongst winged species (Kirschvink and Walker, 1986). Presti and Pettigrew located ferromagnetic deposits in the neck muscles of pigeons and white-crowned sparrows (1980) and found that biogenic magnetic particles were spread uniformly throughout the skull. They also examined 40 other species for inducible remanence, with widely varying levels being found; appreciable levels were associated only with migratory birds. To provide the observed remanence, they estimated that about 6×10^6 aligned, single-domain cubic particles of size 0.07 mm were needed, provided the domains did not interact. They further proposed the torque due to these particles could stimulate a spindle muscle receptor.

In the marine environment, a similar magnification of the LGF seems to occur. Frankel (1986a, 1986b) suggested that certain bacteria and algae could navigate along the LGF due to the torque produced by the LGF and chain-like deposits of magnetite forming natural magnetic dipoles. In the case of the magnetotactic algae, the organism can contain thousands of particles aligned in many collinear chains within the cell.

Electrical methods of amplification are also used to enhance ambient LEFs in the marine environment. The shape of certain fish appears to provide amplification of LEFs in the form of surface fields. Pores located near sites on the surface of the skin of the fish may conduct surface ion flows to the electroreceptor sites. Elasmobranch or cartilaginous fish, the most electrosensitive species known, rays in particular being able to respond to fields of the order 0.5 mV m^{-1} (Adey, 1981), possess what is termed the ampullae of Lorenzini (Murray, 1967; Obara and Bennett, 1972; Fishman, 1987). This system of pores, canals, receptor cells and attached nerve fibres acts as a functional unit to provide the fish with a means of detecting small E fields.

4.6 Simple Viscosity Model for Single Ion Translational Diffusion

A number of theoretical ion diffusion models have been based upon simple viscosity. Durney et al. (1988) used an elementary viscous model to study static and extremely low frequency E and B field exposures. Considering only the Lorentz forces acting upon ions, they studied the stability of the equation of motion via its eigenvalues. Galt et al. extended this work to include a potential well, which models its potential energy (1993). Halle also used a viscous model to estimate the role of viscosity upon possible E and B field effects (1988). Kinouchi et al. (1988) used a random walk viscous model to examine static B field effects, which included terms for both Lorentz forces and Maxwell stresses as applied to ions and magnetically susceptible particles. Unlike the random walk model described in the next section, which has no viscosity term, their model used a microscopic viscosity of 3×10^{14} s^{-1}. Chiabrera and Bianco (1988) used a viscous model to examine the role of the LGF at ligand- (ion-) binding sites. Their model included a random noise term with zero average.

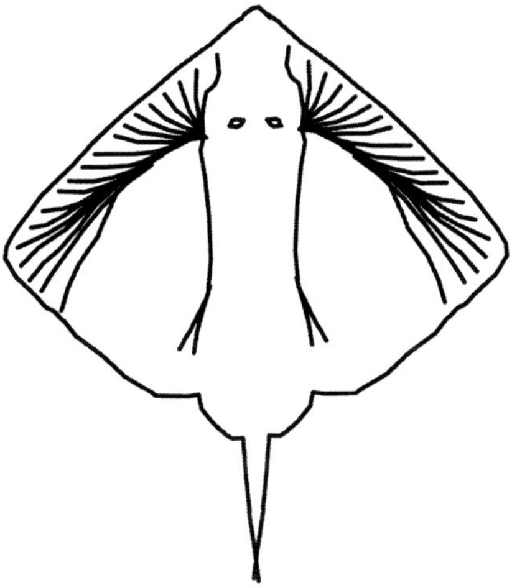

Figure 4.8 Electrosensitivity observed in predatory behaviour of the elasmobranch species. The diagram shows the anatomy of the ampullae of Lorenzini, including the pores at the edge of the ventral sides of the fin, and the network of canals leading to the centrally located ampullae.

In conventional viscous models, time-averaged macroscopic motion is analysed by lumping all collisions together into a very familiar macroscopic 'friction' term (Papoulis, 1965; Reif, 1965). Ionic motion is described by the viscous Lorentz equation where motion is retarded by the drag term, $m\nu_{\upsilon}$ (Durney et al., 1988).

$$m\frac{\mathrm{d}\upsilon}{\mathrm{d}t} = qE + q\upsilon \times B - m\nu_{\upsilon} \qquad (4.3)$$

Relating this equation to the physical world, ionic mobility (velocity per unit E field, $\mu = \frac{\mathrm{V}}{E}$) can be measured experimentally (Pethig, 1988). From Eq. (4.1) it is seen that for the case $E = 1$ V m^{-1} at constant drift velocity ($\frac{\mathrm{d}\upsilon}{\mathrm{d}t} = 0$), a macroscopic viscosity can be defined $\nu = \frac{q}{m\mu}$, ignoring any effects due to the LGF. Using calcium ions as our example, where the ionic mass m is $40 \times 1.67 \times 10^{-27}$ kg, the ionic charge q is $2 \times 1.6 \times 10^{-19}$C, μ_{Ca} is measured to be 6.2×10^{-8} m s^{-1} per V m^{-1}. Hence $\nu = 7.8 \times 10^{13}$ s^{-1}.

Figure 4.8 shows the motion that was analytically calculated from Eq. (4.1), assuming a range of different viscosities with a 1 V m^{-1} E field applied in the x direction. The particle is assumed to start from the origin with non-zero initial velocity ($\upsilon_{\mathrm{tot}} = 200\sqrt{3}$ m

s^{-1}), even though this is a macroscopic model. After a transient response, the particle moves in the x direction with constant velocity corresponding to the mobility. If the initial velocity were zero, time stationary motion is unaffected.

For the case of diffusive effects due a static E field Fig. 4.9 shows the ionic motion in a field of strength 1.0 V m^{-1}. Fig. 4.10 shows the ionic motion in a B field of strength 60 μT, directed along the Z axis at various viscosities. With no viscosity, the particle performs a helical path, a circle in the XY plane. As viscosity increases, the motion becomes a spiral towards a focal point centred off-origin. In other words, the ion moves deterministically to a point in the transverse XY plane parametrically in time (i.e., at time t, the ion is located at point $P(X(t),Y(t))$. As viscosity increases, the distance moved $(O-P)$ in time t is reduced.

At realistic collision rates (where the collision time is given by the inverse of the viscosity), the model predicts that diffusive motion is damped out, except in the case of very large fields. The difference between this motion and that of the E field is due to fact that the B field does no work to move the ion. While the E field opposes the drag term, the B field does not. Assuming non-zero initial velocity, motion ceases after a period, while the particle remains at the origin where the initial velocity is zero.

Macroscopic estimates can be derived as to the level of static B fields that produce observable time-averaged effects. From their eigenvalue study, Durney et al. (1988) concluded that significant B field effects are unlikely due to viscosity. Halle also concluded that effects are swamped by viscosity (1988). Kinouchi et al. estimated from their viscous random walk model that a threshold B field of 10^6 T is necessary to produce a 10% retardation effect upon ions (1988). Again, Chiabrera and Bianco (1987) estimated that an effect might be expected for a B field of more than 10^7 T. Due to the very large size of the estimated B fields involved, no such experimental observations exist of B field effects upon ionic diffusion. Viscous calculations have been successfully compared, however, with experiments involving rotations of magnetically susceptible particles of relatively large size (about 1 μm) (Kinouchi et al., 1988) or where macroscopic drifts are involved (WHO, 1987).

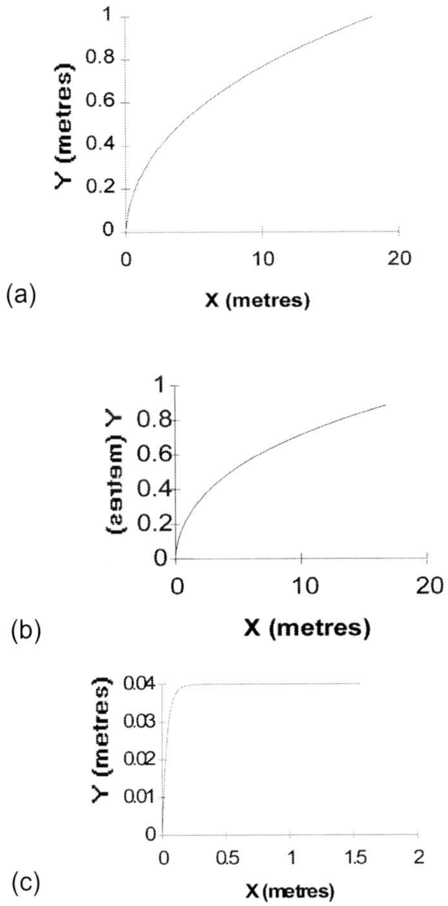

Figure 4.9 (a) Motion of an ion in the XY plane using Eq. (4.1), $E_0 = 1.0$ V m^{-1}, $B_0 = 0$ T, $v = 0$ col s^{-1}. (b) Motion of an ion in the XY plane using Eq. (4.1); $E_0 = 1.0$ V m^{-1}, $B_0 = 0$ T and $\nu = 50$ col s^{-1}. (c) Motion of an ion in the XY plane using Eq. (4.1); $E_0 = 1.0$ V m^{-1}, $B_0 = 0$ T and $\nu = 5{,}000$ col s^{-1}.

The viscous model presented above is limited for the following reasons: (1) there is no mechanism by which the photons interact with the model. In particular, as discussed, photons have their own intrinsic states by which they interact as the binding energy within molecular aggregates; (2) similarly lumping all E and H field effects into those associated with ions and proteins is incomplete because

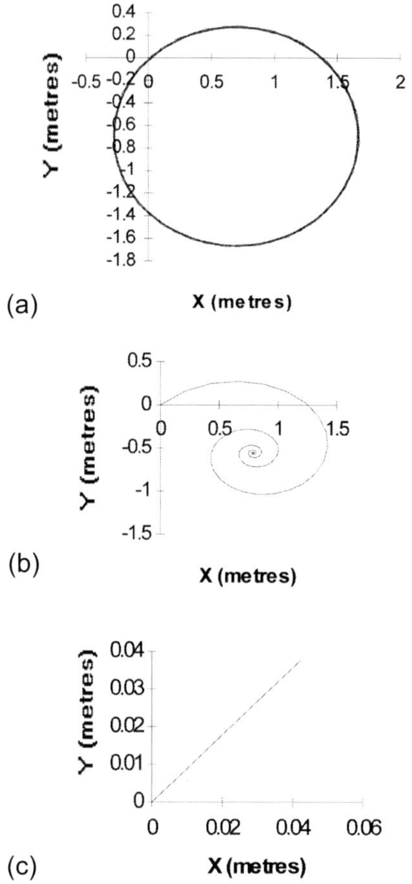

Figure 4.10 (a) Motion of an ion in the XY plane using Eq. (4.1); $E_0 = 0$ V m^{-1}, $B_0 = 60$ T and $\nu = 0$ col s^{-1}. (b) Motion of an ion in the XY plane using Eq. (4.1); $E_0 = 0$ V m^{-1}, $B_0 = 60$ T and $\nu = 50$ col s^{-1}. (c) Motion of an ion in the XY plane using Eq. (4.1); $E_0 = 0$ V m^{-1}, $B_0 = 60$ μT and $\nu = 5{,}000$ col s^{-1}.

photons can by themselves induce macroscopic chemical and molecular effects, not just the other way around. The viscous model assumes that E and H field effects are only associated with charged particles such as electrons and larger-sized particles, yet biophotons are much smaller. There are thus streams of biophotons flowing everywhere inside flows of ions and polar proteins; (3) where large periods or large enough numbers of diffusing charges are involved, as in repairs of wounds (periods of perhaps several days) or cellular division cycles (nine months for human reproduction periods), stochastic diffusion processes involving ions and proteins occur despite opinions to the contrary. We examine the case of protein diffusion later in Chapter 5, where large numbers of cells, each with many membrane proteins and long diffusion periods, need consideration to achieve alignment of cells with latticed (liquid-crystal) membranes to produce magnification of ambient biogenic fields.

4.7 A Translational Random Walk Model for Ion Diffusion

The single-ion model to be presented in this section uses a simple hard-sphere model[a] for the electric charge and the collisions of the ion. A simple magnetic dipole moment is added to study the internal rotation of the ion. Both models use Monte Carlo methods to choose the collision periods, the velocities associated with the thermal and electrical energies and the velocity offsets associated with the magnetic rotational energies. There are other models of ionic diffusion (Allan and Tildesley, 1987; Birdsall, 1991), more detailed than the random walk model that is presented below. Random walk models also suffer the limitations mentioned above in regard viscous models. However, the present model is appropriate as it is contemporary with bioeffect models and reveals magnetic effects that have not to date been reported in the literature.

Figure 4.11a illustrates a Brownian motion random walk process for several collisions, while Fig. 4.11b shows the velocity magnitude. In this case there are no external E or B fields. The model assumes that only an impulsive collision force occurs that acts instantaneously at some point in time to affect a change in particle velocity, unlike, for example, more realistic Leonard–Jones potentials that vary with distance. In the absence of applied fields, no forces are assumed to exist between collisions and hence we have

[a]Compared with the computational sophistication used in current plasma and liquid models, some might call a hard-sphere model primitive, yet it can be a useful computational and theoretical tool of study.

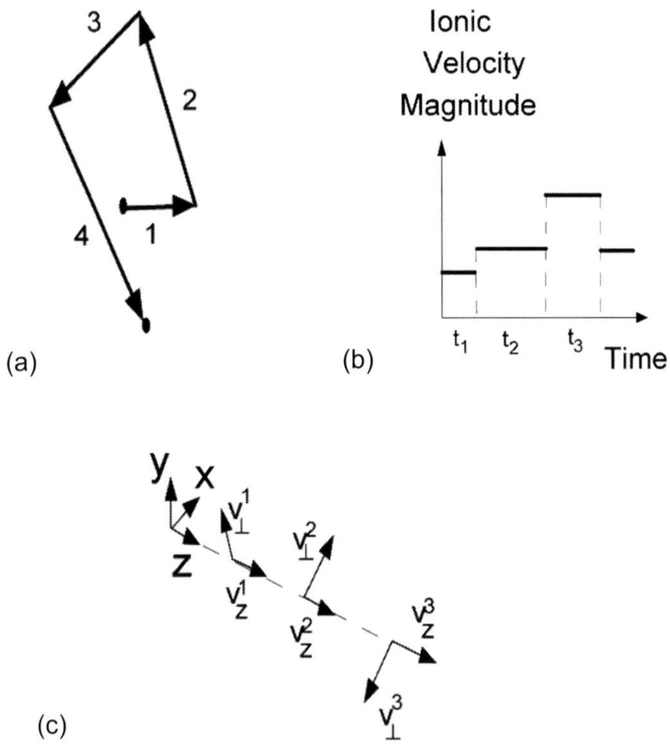

Figure 4.11 Random walk thermal agitation only. (a) Positional offsets, (b) velocity magnitude and (c) velocity parallel and orthogonal to the *z* axis.

a free-flight period. Figure 4.11c shows the velocity components parallel and orthogonal to the z axis of Cartesian (x, y, z) and cylindrical (ρ, ϕ, z) coordinates. The random walk model simplifies the actual physics, ignoring large fluctuations that occur in the collision process due to electrostatic forces between outer shell electrons; the single-ion model considers only short-range effects.[a]

When a uniform, static E field is applied to the medium and ion, the random walk is modified, as illustrated in Fig. 4.12a,b, which shows the position and velocity offsets. Figure 4.12c shows

[a]While the present model treats only the motions of single ions, multi-particle models must consider long-range ion–ion interactions. In bioeffect models, long-range effects are best tied in to the ionic concentrations rather than using the more conventional Ewald sums, etc. (Allan and Tildesley, 1987).

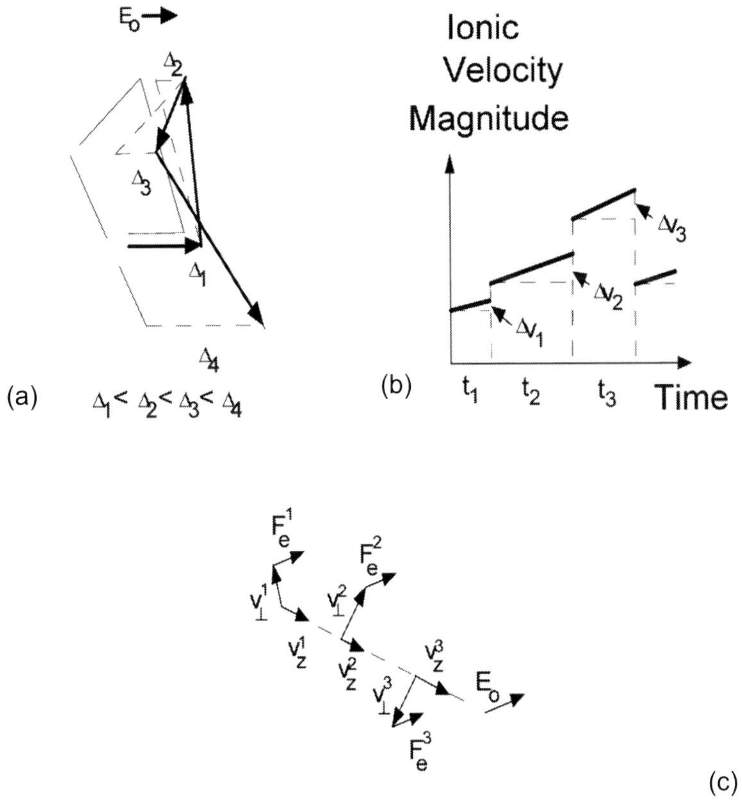

Figure 4.12 Random walk E Field-induced effects. (a) Positional offsets, (b) velocity magnitude and (c) velocity parallel and orthogonal to the z axis.

the velocity components parallel and orthogonal to the z axis and the forces due to the applied E field. Since the offsets in this case are all directed along the applied field, the position offsets accumulate monotonically. Applying a uniform, static B field to the liquid and ion, the effects on the offsets are illustrated in Fig. 4.13a,b. Due to the helical path of the ion, the field produces a curved positional offset in the plane normal to B_0, orthogonal to the velocity at all times, leaving its magnitude unchanged. Unlike the case for a uniform, static E field, the offsets are randomly directed in the plane orthogonal to the B field and vary randomly in direction, accumulating in a different manner to the E field. Figure 4.13c shows the random walk in terms

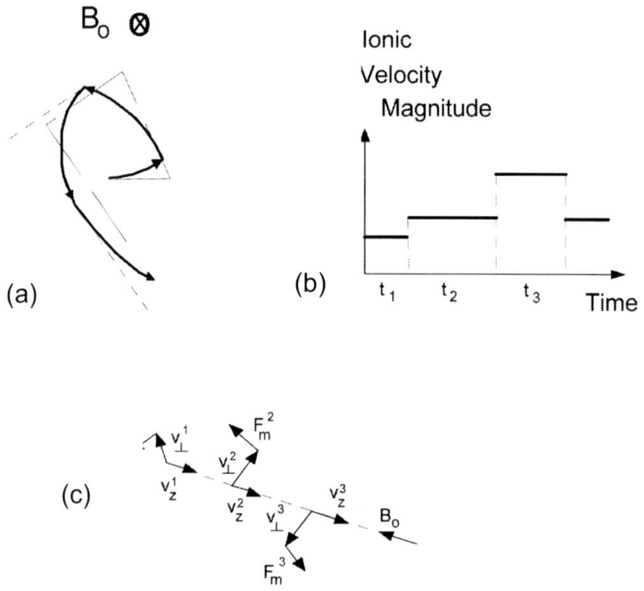

Figure 4.13 Random walk B Field-induced effects. (a) Positional offsets, (b) velocity magnitude and (c) velocity parallel and orthogonal to the *z* axis.

of the velocity and force components parallel and transverse to the applied B field.

Assuming both an E field and a B field is involved, a Lorentz equation for the motion of the ion between collisions may be written where, like other random walk models (Reif, 1965), no viscous term exists.

$$m\frac{dv}{dt} = qE + qv \times B \tag{4.4}$$

Whereas the viscous model lumps the entire collision process into a viscous term, the random walk model inserts each collision explicitly. The walk is constructed by generating random thermal velocities after each collision. Random numbers allocate the thermal velocities that have a Gaussian probability distribution. The collision periods having a Poisson probability distribution are also randomly selected, the user specifying the average collision rate. The Lorentz equations are solved analytically to yield the current position after this period, and a collision is then assumed to occur. The process is now repeated until the desired total time has elapsed. The result is a very simple simulation for the motion of a single ion in a liquid (Allen and Tildesley, 1987).

Examining the random walk process in detail, the numerical solution is updated after each collision and a new ionic velocity is probabilistically determined in the following manner. First the time domain $t_{\max}$ is divided into a number of collisions, N_c, and associated periods t_i of random duration. Thus

$$t_{\max} = \sum_{i=1}^{N_c} t_i \tag{4.5}$$

and the mean collision period

$$\bar{t} = \frac{t_{\max}}{N_c}. \tag{4.6}$$

A uniformly random variable p_i^{col} is used to determine the collision period, which has a Poisson probability distribution. The average period is governed by the average collision rate $(1/\bar{t})$ for the process; for a calcium ion $1/\bar{t} \approx 10^{14}$ col s^{-1}, depending on the temperature and pressure. The individual time intervals are

$$t_i = -\bar{t}\ln(p_i^{\text{col}}) \tag{4.7}$$

From the Maxwellian thermal distribution (Reif, 1965), a probabilistic thermal velocity is selected, initially, and after each collision at time t, using a Gaussian distribution of the variables q_i^{col}, r_i^{col} and s_i^{col}.

$$v_i^x(t) = q_i^{\text{col}}\sqrt{2k_B T/m} \tag{4.8a}$$

$$v_i^y(t) = r_i^{\text{col}}\sqrt{2k_B T/m} \tag{4.8b}$$

$$v_i^z(t) = s_i^{\text{col}}\sqrt{2k_B T/m} \tag{4.8c}$$

Due to increases in velocity gained from the E field between collisions, small corrections are made to the velocity distribution at each collision (Kittel, 1958). Thus the random velocities have a 'memory' of the previous state of the system (Reif, 1965). The B field does not modify this distribution since it does not affect translational velocities and energies. It does, however, modify the rotational energies (Fig. 4.14).

According to Reif, the classical approximation used above to model ionic diffusion is valid if dimensions of the model are larger

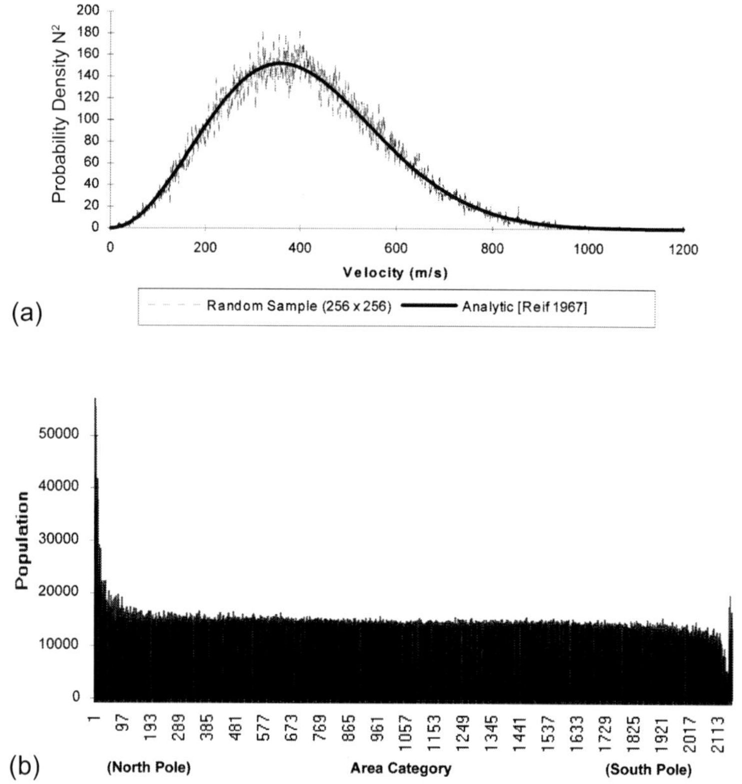

Figure 4.14 (a) Maxwell velocity distribution compared to analytic distribution (Rief, 1967). (b) Distribution of directions associated with velocities over a spherical surface using a 1° grid. Visible is the round-down error associated with counting over finite areas.

than the de Broglie wavelength ($\lambda_0 = \frac{\hbar}{p_0}$; p_0 is the momentum), otherwise a quantum description is needed (1965). In this case, $\lambda_0 = O(10^{-12})$m. The mean free path of the model is also $O(10^{-12})$m, so the model is marginal. We are at the interface between quantum and classical.

The random walk model described above allows investigation of the complete translational motion of ions, including the stochastic components. Using physical parameters suitable for calcium ions in solution at 300 K, the motion between collisions was computed for a range of E and B field exposures. Initially the motion for single steps (of the random walk) was studied to check the numerical model. In these single-step cases the results could be tested using explicit solutions for the transverse ionic motion (x, v_x, y and v_y) when either an E field is directed along the x axis or a B field is directed along the z axis. The complete random walk model without the applied E or B fields was then tested to check various velocity parameters, including its predicted average thermal velocity for calcium ion diffusion. A series of computations ensued, each involving at least 65,636 collisions and using a collision rate of 7.8×10^{13} col s^{-1}. Using the results for many collisions, the mean velocity in the X direction was estimated to be 203 m s^{-1}, reproducing well the parameters of the underlying theory. At 300 K, the mean thermal velocity of the ion may be estimated to be 203 m s^{-1}, where 0.5 m $< v_{\text{th}}^2 >= kT$, and k is the Boltzmann constant (Reif, 1965). Other tests that gave confidence in the random walk were conducted using the analytic solution for the offset due to a static B field undergoing a series of collisions.

The complete model was used to test the influence of static E and B fields. Using an E field of 1 V m^{-1} in the X direction, the time evolution for the total cylindrical displacement, ρ, is shown in Fig. 4.15. Again, with a Z-directed B field of 1.0 T, the result is also shown in Fig. 4.15. The collision rate, 10^6 col s^{-1}, was chosen to

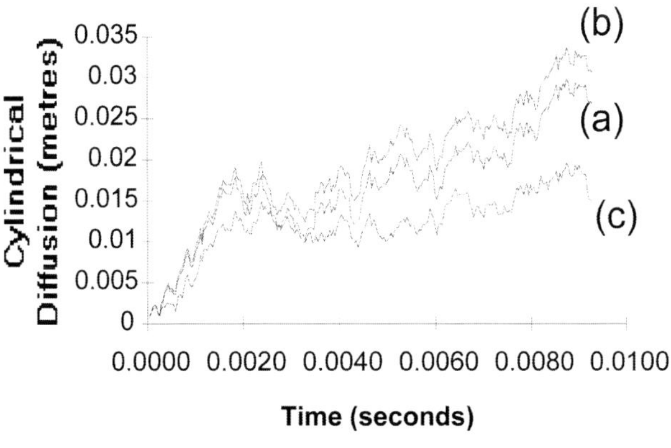

Figure 4.15 Total cylindrical diffusion, ρ, in the XY plane for (a) unperturbed, (b) E field- and (c) B field-influenced random walks (collision rate 1.0×10^6 col s^{-1}).

visibly demonstrate effects of these fields upon the thermal motion using the same random collision sequence, which is also shown for comparison. The E field in this case increases the cylindrical diffusion rate; the B field reduces it.

Another comparison was performed between the uninfluenced thermal motion and static E and B field exposure levels using a physically realistic collision rate 7.8×10^{13} col s^{-1}. No discernible difference between the uninfluenced thermal agitation and the E and B field cases is now seen by simple visual inspection. Since the offsets are known to be orders of magnitude less than the thermal velocities, the results are no surprise. However, such tiny offsets have macroscopic effects when the averaging time is long enough. These results when examined carefully illustrate how in liquids the offsets induced by E and B fields are orders of magnitude less than the thermal velocities. Yet in the macroscopic world, these offsets are clearly visible as ionic currents, while thermal noise may not be as apparent. We must be careful not to ignore macroscopically significant offsets because of their apparent insignificance with thermal motions when viewed microscopically.

Numerical examination of the offsets yields insight into the collision sequences. Figure 4.16a shows the total cylindrical offset ρ for the E field case, while Fig. 4.16b shows the distribution of the phase angle ϕ of the offset directions where the X direction corresponds to zero phase angle ($\phi = \arctan(y/x)$). Similarly for the B field case, Fig. 4.16g shows the total offset r, while Fig. 4.16h shows a distribution of the offset phase angle ϕ. The angles of the thermal trajectories for both E and B field cases are not shown but are randomly distributed similar to the B field offset phase angle distribution shown in Fig. 4.16h. Distributions of the angular difference between thermal and offset directions are shown in Figs. 4.16c and 4.16i for both cases. These plots show the coherence of the static E field offsets and the stochastic nature of the B field offsets, while demonstrating the relationships between the directions of the offsets and the thermal motion. It is seen that depending on whether the offsets are coherent or stochastic, there is a dramatic effect upon the total offset. The presence of a time invariant ionic mobility (6.2×10^{-8} m s^{-1} per V m^{-1}) is seen in Fig. 4.16a, while a similar drift is lacking from Fig. 4.16g. These results are further discussed

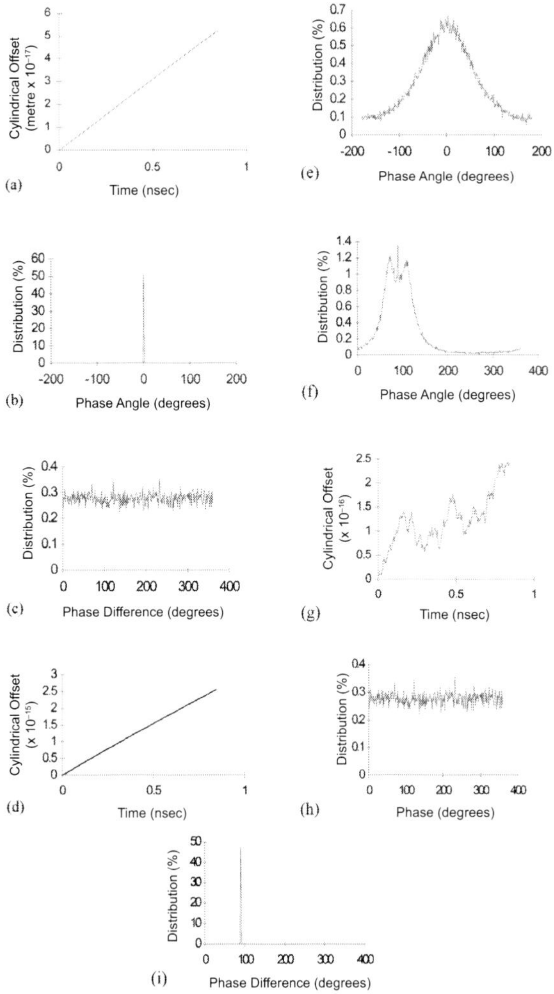

Figure 4.16 (a) Cylindrical diffusion vs. time, where $E_x = 1$ V m^{-1} and $B_0 = 0$ T. (b) Distribution of offset phase angles, where $E_x = 1$ V m^{-1} and $B_0 = 0$ T. (c) Distribution of phase differences between thermal and offset angles, where $E_x = 1$ V m^{-1} and $B_0 = 0$ T. (d) Cylindrical offset vs. time, where $E_x = 50$ V m^{-1} and $B_0 = 0.25$ T. (e) Distribution of offset phase angles, where $E_x = 50$ V m^{-1} and $B_0 = 0.25$ T. (f) Distribution of phase differences between thermal and offset angles, where $E_x = 50$ V m^{-1} and $B_0 = 0.25$ T. (g) Cylindrical diffusion vs. time, where $E_x = 0$ V m^{-1} and $B_0 = 1$ T. (h) Distribution of offset phase angles, where $E_x = 0$ V m^{-1} and $B_0 = 1$ T. (i) Phase between thermal and offset angles, where $E_x = 0$ V m^{-1} and $B_0 = 1$ T.

in the following sections where viscosity is again studied. One more case is shown in Fig. 4.16d–f. Here, both static E and B fields are present and the ratio between the E and B field forces is close to balance ($E_0/v_x B_0 = 1$). Now, the offsets may be viewed as a fixed vector plus a rotating vector of equal length. In the particular case of Fig. 4.16d, the mobility is reduced by around 10% by the stochastic B field offsets.

This mobility effect was further investigated. A longer time interval was now used as the B field was varied between 0 and 0.1 T and the E field was kept at 1 V m^{-1}. Figure 4.17a,b shows results for the cases where the B field was 0 T and 0.1 T using $t_{max} = 840$ ns. As seen, the total offset remains relatively constant over the entire time period where the B field is not present, while it becomes very noisy as the B field is increased to 0.1 T. In statistical terms, the mean total offset remains constant, while its variance (σ) increases dramatically as the B field varies (Fig. 4.17c). In comparison with the level of the stochastic motion associated with the random walk, however, this effect is still minuscule. What is unclear from this analysis is how the total offsets may be treated by biological receptors, given that the B field offsets are non-linear in time unlike the random walks, which are linear.

Figure 4.18a,b shows the results for two cases: (a) $B_z = 0$ T, $E_x =$ 1 V m^{-1}, and (b) $B_z = 1$ T, $E_x = 0$ V m^{-1}. Clearly the particle orbits in the presence of the B field but not when the E field is present. The angular frequency for the case shown in Fig. 4.18b was found to be half the corresponding cyclotron frequency.

4.8 Analysis of Diffusive Offsets Due to Static E Fields

Comparing results it is found that after a transient initial response the viscous and random walk models agree; E field effects can be described satisfactorily by the random walk and viscous models because the offsets are not stochastic in direction. As well, the mobility of ions due to E field exposure stems from relatively minuscule offsets to the unperturbed thermal effects. In this latter case, extremely large E field exposure levels are often considered necessary to cause observable effects above thermal

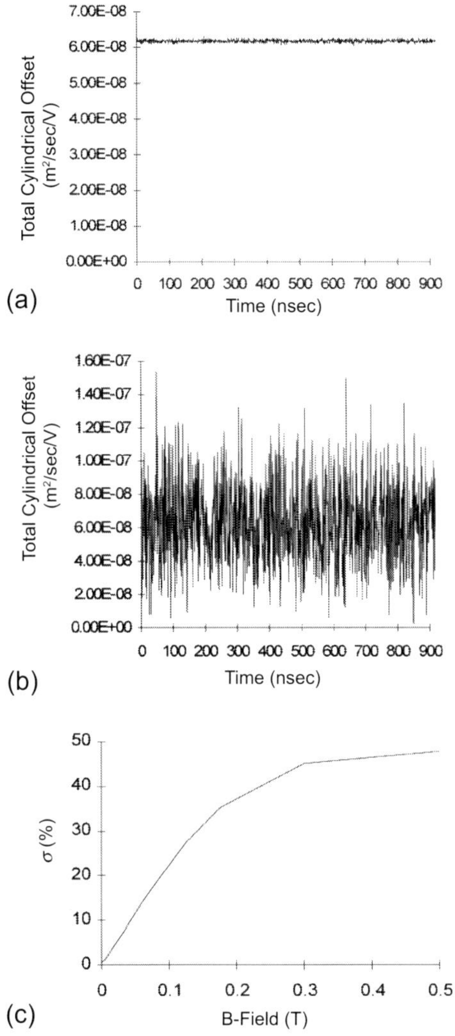

Figure 4.17 (a) Total cylindrical offset vs. time, where $E_x = 1$ V m^{-1} and $B_0 = 0$ T. (b) Total cylindrical offset vs. time, where $E_x = 1$ V m^{-1} and $B_0 = 0.1$ T. (c) Variance of total cylindrical offset vs. B field, where $E_x = 1$ V m^{-1}.

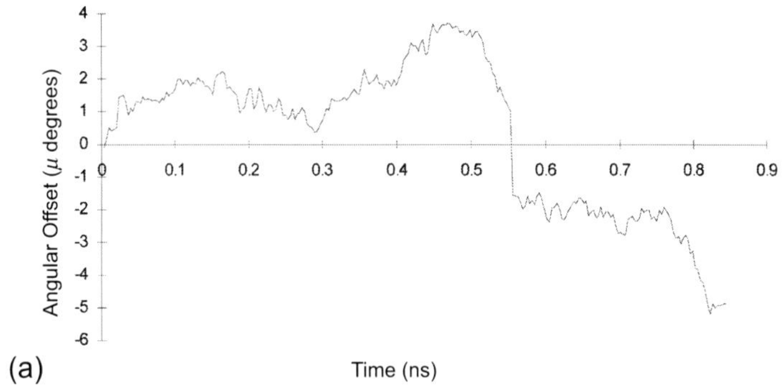

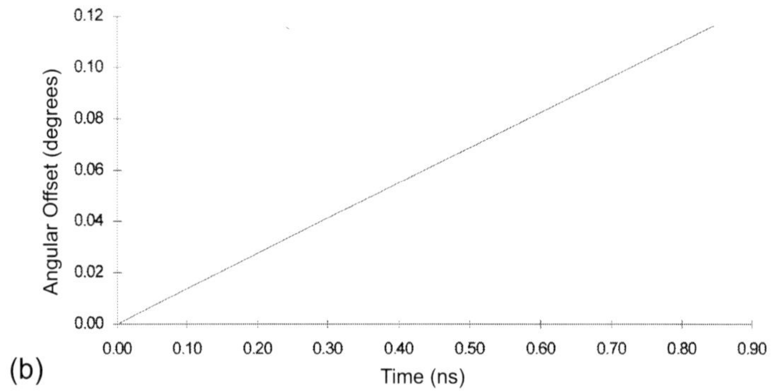

Figure 4.18 (a) Orbital diffusion vs. time, where $B_z = 0$ T and $E_x = 1$ V m^{-1}. (b) Orbital diffusion vs. time, where $B_z = 1$ T and $E_x = 0$ V m^{-1}.

agitation (Adair, 1991a). Mobility studies show, however, that tiny, non-random microscopic drift velocities are observable at the macroscopic level. Barnes (1986) reported drift velocities for sodium ions in solution 10^{-10} times that of its mean thermal velocity, while Ciccotti and Jacucci (1975) reported similar findings for liquid argon at low temperatures.

Calcium ions may be examined using a hard-sphere model. Figure 4.11a,b shows that the E field induces an acceleration between collisions, which results in a small drift velocity and associated positional offset to the thermal walk. The final velocity offset after one collision period is qEt/m, where m is the mass, q the ion charge, E the amplitude of the E field and t the collision period, while the average drift velocity during this collision period is half this final value (Reif, 1967, in particular p. 344). From experimental data (μ_{Ca}), the mean collision time is estimated to be 1.3×10^{-14} seconds. This value is consistent with simple viscous theory where $\nu = 7.8 \times 10^{13}\ \mathrm{s}^{-1}$ (and is also consistent with the hard-sphere model of viscosity (Chiabrera and Bianco, 1987).

These offsets can be theoretically analysed. The E field offsets to the random thermal diffusion are non-randomly directed and accumulate monotonically, as illustrated by Fig. 4.11a. The intercollision spacing, d_i, can be written in terms of the microscopic average velocity between collisions (see Fig. 4.11a–c).

$$d_i = \frac{|\vec{v}_{i+1} - \vec{v}_i|}{2} t_i = \left[|\vec{v}_i| + \frac{\Delta v_i}{2} \right] t_i \tag{4.9}$$

where $\vec{\mathrm{v}}_i$ and $\vec{\mathrm{v}}_{i+1}$ are the initial and final velocities, Δv_i the velocity offset and t_i the collision period. It is seen there is both a random and a non-random offset.

Microscopically, the velocity offsets will be swamped by thermal velocities, which are associated with random E field fluctuations, otherwise known as the Johnson–Nyquist noise potentials. In accordance with Coulomb's law, these noise fields originate from the random space-charge distributions inside the solvent. The Nyquist theorem relating these fluctuating potentials $\bar{v}_n$ to a resistance R allows their measurement, $\bar{v}_n^2 = 4\mathrm{k}T R\Delta f$ (Reif, 1965), where T is the absolute temperature, k is the Boltzmann constant and Δf is the bandwidth of the noise. Macroscopically, after many collisions, the non-random positional drift due to the velocity offsets may dominate any Fick's diffusion associated with the much larger thermal agitation.

Summing over many collisions, and assuming the random terms due to thermal drift are macroscopically negligible, the macroscopic mobility is written in terms of the average velocity offset and time interval.

$$\mu_{\mathrm{Ca}} = \frac{1}{E_0} \frac{1}{N_{\mathrm{c}}} \sum_{i=1}^{N_{\mathrm{c}}} \Delta v_i = \frac{q}{m} \frac{1}{T} \sum_{i=1}^{N_{\mathrm{c}}} t_i \tag{4.10}$$

where N_{c} is the number of collisions, each individual collision period is t_i and the total time interval is T. If the thermal velocity of an ion is assumed to be aligned with the x axis after collision, $\vec{v}(t) = v_{x0}\hat{x}$, and if there were no external forces present, the mean collision time, the time taken for the particle to travel the mean free path distance d would be $t_{\text{col}} v_{x0}$. The modified time taken between collisions, t_m, for an ion moving under the influence of a uniform static E field aligned with the x axis, can be obtained analytically assuming t_{m} and $t_{\text{col}} << 1$.

$$\frac{q}{2m}\frac{E_0}{v_{x0}}t_{\text{m}}^2 + t_{\text{m}} - t_{\text{col}} = 0 \tag{4.11a}$$

and the change in position can be described as

$$x = v_{x0}t_{\text{col}} + \frac{1}{2}\frac{qE_x}{m}t_{\text{col}}^2 \tag{4.11b}$$

For the case of mobility due to a 1 V m^{-1} E field, the relative change in time over a single step is equal to the ratio of mean velocity offset to mean thermal velocity, 3.1×10^{-10}, ratios similar to those found by Barnes (1986) and Ciccotti and Jacucci (1975). The much larger noise fields are not able to drown out the ion mobility where enough time or sufficient field levels exist. Likewise, the positional offset during one collision can be estimated analytically to be a tiny 4.0×10^{-22} m. The random walk model was checked using a single step of the walk with a 1 V m^{-1} E field in the X direction. The positional offset was numerically calculated to be 3.92×10^{-22} m, comparing well with the analytic result. Note that this positional offset is tiny even compared with the dimensions of the ion (10^{-10}m). We picture a very crowded collision process that takes many collisions to move substantial distances due to the accumulation of offsets of a 1 V m^{-1} E field. Such tiny distances are not unique to the random walk model but are also associated with the viscous model; experimental mobility rates also indicate that positional drift is very tiny when taken over one collision period.

It can be seen from Fig. 4.16a,b, in regard to ensemble effects, that E field offsets are directed along the applied field. For a uniform spatial distribution of ionic particles, as exists under equilibrium conditions, the offsets from individual ions sum together to affect equilibrium. The same applies to non-equilibrium conditions. These effects are related to the ability of the E field to do work as motion proceeds.

According to CEM, static E fields require physical contact with the solvent as only conduction currents are involved, compared with oscillating fields where displacement currents also flow. Biologically, such contact for humans is abnormal except in biomedical applications (and perhaps tactile sensitivities). Yet ionic flows due to E fields are a natural part of the marine environment. Bennett and Obara (1986) investigated the sensitivity of the elasmobranch species of fish known to be able to detect E fields as low as 0.5 μV m^{-1}. This tiny E field can be expressed as a flux of particles $j_E = \mu \vec{E} C$, where j_E is the flux due to an E field acting upon a concentration C of ions in solution (Plonsey and Barr, 1988). For $C =$ 1 M, j_E is estimated to be 6.2×10^{-14} M m^{-2} s^{-1}, equating to a flux of 3.8×10^{10} particles m^{-2} s^{-1}, 38,000 particles mm^{-2} s^{-1}, seemingly a tiny proportion of the total number of ions (1.2×10^{15}) in a 1 M solution of volume 1 mm^3. While the microscopic Johnson–Nyquist noise fields due to random collisions are huge by comparison with such tiny fields, macroscopically these noise fields are not able to drown the biologically detectable minuscule flux of ionic particles due to the accumulation of very many, tiny offsets.

How then are such tiny E fields able to be detected? Are natural magnification processes at work to make elasmobranch fish more sensitive than thermal noise? Certainly Fishman (1987) suggests that the output from a large number of receptor cells in the ampullae of Lorenzini sum together, acting in parallel fashion. In Chapter 5 a hypothetical SFT magnification process involved in fertilisation across species is studied. On the basis of the random walk model, it is seen below elasmobranch electrosensitivity is below the thermal estimates of detectable E fields.

4.9 Estimate of Lowest Detectable E Field by Elasmobranch Fish

An important aspect of E field exposure is to consider the lowest levels that are biologically significant. For E fields to be biologically significant, they must lead to outcomes significantly greater than the ordinary thermal interactions of the particles with their environment (Adair, 1991a). According to Nyquist's theorem, there is a bandwidth, or equivalently an averaging time, over which thermal noise and the non-random E or B fields are compared (Barnes and Seyed-Madani, 1987). The volume (or spatial dimension) of comparative measurements must also be taken into account since for small volumes, statistical fluctuations due to space charge density become relatively larger. For example, at the molecular level, noise fields are of the order of 10^8 V m^{-1}. In the following, the effects of both random noise fields and non-random applied fields are simply compared, rather than comparing the fields themselves.

The x component of the translational Lorentz equation of motion is written as $x = v_{x0}t_{\text{col}} + \frac{1}{2}\frac{qE_x}{m}t_{\text{col}}^2$, where the x component of the thermal velocity is v_{x0}, and t_{col} is the time since the last collision. Hence the thermal step is $v_{x0}t_{\text{col}}$, which is random, while the offset is $\frac{1}{2}\frac{qE_x}{m}t_{\text{col}}^2$, which is non-random.

At this point we need to decide upon a suitable averaging time and volume size for our comparison of the effects due to noise and due to the E fields of interest. The averaging time in the general case relates to the method of measurement, whether man-made experimental equipment, or perhaps some biological means of sensing E fields. Our measuring volume will also depend on the measurement technique, whether experimental or biological. We can estimate that the averaging time (t_{ave}) might range from some milliseconds upwards, and our volume will have a length associated with the volume (l_{vol}) that might range from atomic dimensions, 0.1 nm upwards. When considering E field effects upon a calcium ion, the ion's mobility, μ_{Ca}, links the averaging time t_{ave}, with the measurement length l_{vol}; if we choose either t_{ave} or l_{vol}, we specify the other. To determine the volume, a measurement area must be specified; if we are investigating E field effects, this area will form a current probe, sensing a flow of charged particles per area, a current density. Like l_{vol}, this area might range from 0.1 nm upwards, depending on current density.

Taking $t_{\text{ave}} = 0.1$ seconds, we shall assume a very dilute solution of calcium ions; ion–ion interactions are ignored in our simple study. Examining the size of the thermal step, $v_{x0}t_{\text{col}}$, this is estimated to be 2.6×10^{-12} m and the offset $\frac{1}{2}\frac{qE_x}{m}t_{col}^2$ estimated to be 4.0×10^{-22} m, where we take E_x to be 1 V m^{-1} and the mean thermal velocity is

203 m s^{-1} as calculated previously. Such a small offset indicates how many collisions are required to move significant distances at the molecular level. While the offset is tiny compared with the random thermal step across these time intervals, we are now in a position to estimate how much time is needed before the non-random offsets accumulate to give an overall offset significant when compared with the thermal effect. The ion will be offset on average by 6.2×10^{-9} m in 0.1 seconds. This distance is about 1,650 times larger than a single step, which is significant, and we may thus assess that an E field of 1 V m^{-1} will produce an observable effect if we have a current probe of sufficient area to measure it. Similarly, an E field of 1/1,650 V m^{-1}, or 6.1×10^{-4} V m^{-1} will produce an offset as large as the thermal step size after 0.1 seconds.

For elasmobranch electrosensitivity, in particular a sting-ray pursuing prey (who transmits a biogenic E field by firing its muscles in an attempt to evade capture), we assume a glide speed through the water of 5–10 m s^{-1} and a prey size of 0.5 m, so the averaging time is no more than 0.1 seconds, perhaps 1 second, given that the ion flow could spread out somewhat over a larger volume than the original prey size. Using the estimate obtained above for $t_{\text{ave}} = 0.1$ seconds, we obtain the lowest possible E field that could be detected as 6.1×10^{-4} V m^{-1}.

Applying the theory derived by Barnes and Seyed-Madani (1987) we may write an expression for this same E field 'noise floor' as

$$< E_n^2 >= \frac{4\text{k}T\,\Delta f}{\sigma\,\text{Vol}} \tag{4.12}$$

where σ is the conductivity, and Vol is a measuring volume. For a calcium ion, we need to sense distances of 6.2×10^{-8} m per unit E field, so Vol is dependent upon the E field. The area of the probe volume is equal to that of the pore on the surface of the elasmobranch's skin, about 1 mm^2 (Fishman, 1987). For a conductivity of 0.5 S m^{-1}, (Barnes and Seyed-Madani, 1987), and Δf 10.0 s^{-1}, the E field is 1.7×10^{-4} V m^{-1}, 35% of the random walk estimate. Both estimates incorporate the Boltzmann constant, itself a limit on noise, in very different fashions.

To complete this analysis, an estimate of thermal diffusion will enable another measure of the biological significance of the E field offsets. The Fick's diffusion constant, D, and the associated particle flux, j_{F}, can be written as

$$D = \frac{\mu \mathrm{k} T}{Z_{\mathrm{Ca}}} \tag{4.13}$$

$$j_{\mathrm{F}} = -D\nabla C \tag{4.14}$$

where Z_{Ca} (=2), is the charge of calcium ion and C is its concentration gradient, $\nabla C = \frac{\partial}{\partial \rho}\hat{\rho} + \frac{\partial}{\partial z}\hat{z}$. D is estimated to be 8.0 $\times\ 10^{-10}$ m^2 s^{-1}, while for $C = 1$ M m^{-1}, j_{F} is estimated to be 8.0 $\times 10^{-10}$M m^{-1} s^{-1}. Where ionic concentration equals its gradient, flux due to Fick's diffusion, j_{F}, is greater than the X-directed flux due to E field mobility, j_{E}, for E fields below 2.58 $\times 10^{-2}$ V m^{-1} (Plonsey and Barr, 1988).

Our two estimates notwithstanding, the ability of sharks and sting-rays to sense ultra-weak E fields as small as 0.5 $\times$ 10^{-6} V m^{-1} is experimentally undisputed (Kalmijn, 1981). Given that such ambient E field levels will produce accumulated offsets well below the thermal effect over the estimated averaging time and volume, then the only sensible conclusion is that such creatures must have an innate ability to magnify the ambient E fields. Fishman suggests summation of the output of the large numbers of cell receptors in the ampullae (Fig. 4.19) (1987).

Another method by which magnification of ambient E fields could be achieved is in the geometric shape of the fins themselves. Such fins have very slender profiles, and the edges separating the upper and lower parts of these fins create a sharp media discontinuity. Depending on the 'thinness ratio' near the edge of the fins, a concept discussed in Chapter 4 with regard to metallic implants (in antenna theory, the 'thinness ratio' is the ratio between height and wire radius), incident E fields are enhanced at locations directly adjacent to the body surface, near the ampullae pores. Chen et al. (1986) used a moment method employing 424 surface patches to calculate the surface fields and body currents of a realistically shaped human phantom model exposed to ELF fields. They calculated enhancement factors of 50 at the tips of the outstretched fingers (forming a crucifix posture), akin to the outstretched predatorial elasmobranch fish. The difference between

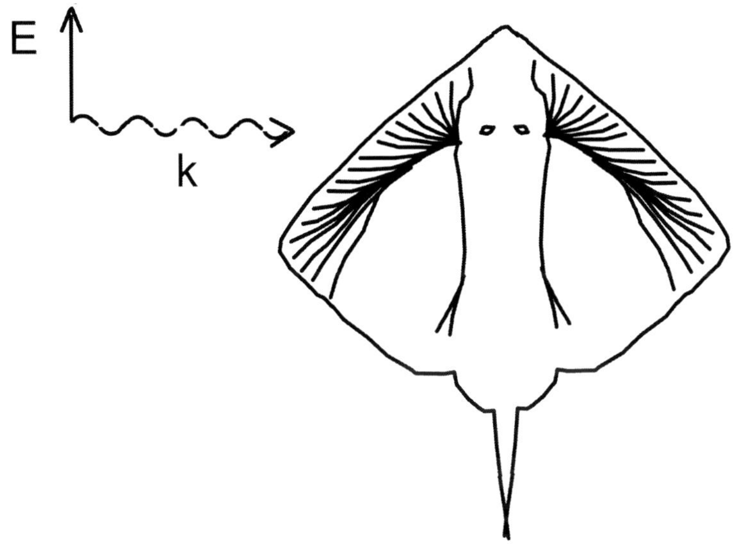

Figure 4.19 E field incident upon sting-ray. Pores are located at the edge of the ventral (and dorsal) sides of the fin, each leading into the network of canals and the centrally located ampullae of Lorenzini.

elasmobranch and humans is in the ratio of the constitutive parameters of the water to that of the fin as compared with the air to the skin of the human body.

It is noted that rays, which have increased electrosensitivity above other elasmobranch fish, themselves the most electrosensitive species currently known, have a very distinctive and pronounced geometry that has a larger thinness ratio at the tips than other elasmobranches, such as the shark. It is also noted that a large proportion of the ampullae are located very close to the fin tips, and coverage is quite extensive to the front of the ray rather than at the rear (Fig. 4.19), indicative of a forward-seeking electrolocating predatory function (Murray, 1967). This is also the case for sharks. The geometric factor for rays and the location of the ampullae pores near the fin edges lend support to shape being a factor in electrosensitivity.

4.10 Analysis of Diffusive Offsets Due to Static B Fields

While viscous theory is consistent with random walk theory developed above for the E field, differences relating to the stochastic nature of the random walk emerge when B field results are compared for the two models. This is due to the random directions of the B field offsets compared with the unidirectional E field offsets. There is also a further difference in the orbital and rotational effects between models.

Where a B field, B_0, is directed along the z axis, and the ion moves in the direction of the x axis following collision, the Lorentz equation can be solved to yield the helical motion illustrated in Fig. 4.10a.

$$x = \frac{v_{x0}}{\omega_c} \sin(\omega_c t) \tag{4.15a}$$

$$y = \frac{v_{x0}}{\omega_c} (1 - \cos(\omega_c t)) \tag{4.15b}$$

where $\omega_c = qB_0/m$ is the cyclotron frequency. For small t ($t << 1$), we obtain an approximation suitable for the steps of the random walk.

$$x = v_{x0} t \tag{4.16a}$$

$$y = \frac{1}{2} \frac{q v_{x0} B_z}{m} t^2 \tag{4.16b}$$

This demonstrates that for a B field exposure, the effect (the force and its consequences) is always orthogonal to the velocity and so no work or diffusive effect on equilibrium conditions can occur.

Similar to the preliminary test used for the static E field random walk, a single-step numerical test was performed using a static B field of 60 μT directed along the z direction and an initial velocity of 200 m s^{-1} in the x direction. Here, the displacement in the y direction was numerically determined to be 4.94×10^{-24} m, comparing well with the analytic result 4.80×10^{-24} m. This simple single-step test is of course only a single time step of the overall random walk. What was required was to see how a sequence of collisions affects the cyclotron motion.

Unlike the viscous model, the random walk model predicts that in the transverse plane, due to the piece-wise helical motion between

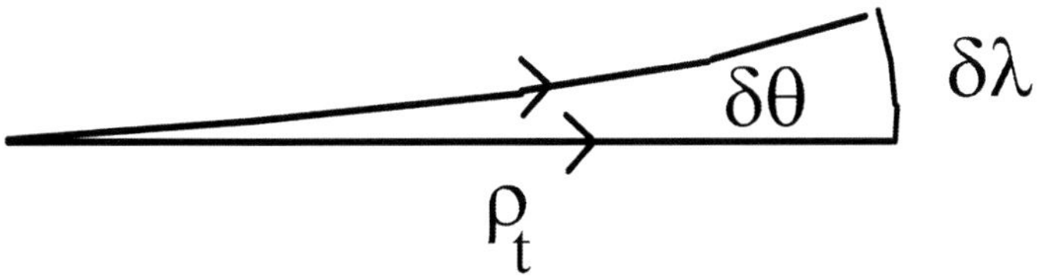

Figure 4.20 Anti-clockwise orbital offset to motion along the x direction due to piece-wise helical motion.

collisions, there is (a) a radially directed but random perturbation to the straight line of the unperturbed path that retards the walk in the ρ direction and (b) a unidirectional but discontinuous orbital (θ) offset. Without a B field present there are no random orbital offsets to the direction of motion.

Figure 4.20 illustrates a single piece-wise helical path between collisions. There is an angular change $\delta\theta$ due to each collision and an associated orbital offset $\delta\lambda = \rho_t\delta\theta$ where ρ_t is the mean free path length in the transverse plane. The initial θ directions are random in accordance with the random collisions, so the unperturbed effect is discontinuous. Since the offset directions (clockwise or anti-clockwise) are dependent upon the direction of the applied static B field, they are unidirectional. When these orbital offsets are averaged over a sufficiently long period, they result in an ionic orbit, analogous to accumulation of offsets due to a static E field, resulting in ionic mobility. The θ offsets have a radius of orbit equal to the mean free path length. For ions (leaving aside larger sized particles, which may be experimentally observable), these θ offsets do not produce a directly observable effect. On the other hand, the ρ offsets can produce a macroscopic effect upon Fick's diffusion in the transverse plane depending upon the level of the B field.

It is seen from Fig. 4.16h that the B field effect is isotropic in the transverse plane since the directions are uniformly distributed in this plane. Thus no retarding effect should be observed where an ensemble of ions exists in equilibrium, since the individual radial offsets counter each other if there is a uniform spatial distribution of ionic particles. However, where a concentration gradient exists, for example, near a source of ions, there will be an ensemble effect, since the ρ offsets do not cancel each other but add. Conversely, there always exists an orbital effect regardless of particle distribution in ensembles.

Considering the helical motion given by Eq. (4.15a–b) the collision period for the ion to transverse the mean free path is modified, increasing[a] with an increasing B field, compared with the unperturbed path. This is obtained by considering the radial displacement in the XY plane and using Eq. (4.15a–b) to solve for the modified time t_m.

$$t_{\mathrm{m}} = \frac{1}{\omega_{\mathrm{c}}}\left(\pi - ar\cos(\frac{1}{2}\omega_c^2 t_{col}^2 - 1\right) \tag{4.17}$$

Expanding the perturbed collision period as a Taylor series where $\omega_{\mathrm{c}} t_{\mathrm{col}} << 1$, we get

$$t_{\mathrm{m}} = t_{\mathrm{col}} + \frac{\omega_{\mathrm{c}}^2 t_{\mathrm{col}}^3}{24} + \frac{3\omega_{\mathrm{c}}^4 t_{\mathrm{col}}^5}{640} + \frac{5\omega_{\mathrm{c}}^6 t_{\mathrm{col}}^7}{7168} + \ldots \tag{4.18}$$

Analogous to free-space cyclotron motion (Chen, 1974), and similar to the findings of Kinouchi et al. (1988), the ρ offsets of the B field are thus found to retard diffusion in the transverse plane. This effect modifies the Nernst–Planck equation relating ionic flux to Fick's diffusion and E field mobility (Plonsey and Barr, 1988).

$$j_{\mathrm{T}} = -D\nabla C + C\mu_{\mathrm{Ca}}\vec{E} \tag{4.19}$$

since in the hard-sphere model Fick's constant is dependent upon the mean ionic total velocity, v_{th}, and the mean free path, d: D $= v_{\mathrm{th}} d/3$ (Reif, 1965; Plonsey and Barr, 1988). The total particle flux, j_{D}, due to both Fick's diffusion, j_{F}, and the retardation effect of the B field, j_{B}, is written as $j_{\mathrm{D}} = j_{\mathrm{F}} + j_{\mathrm{B}}$ where

$$j_{\mathrm{B}} = D\frac{\mu_{\mathrm{Ca}}^2 B^2}{24}\nabla C \bullet \hat{B} \times \hat{n} \tag{4.20}$$

where D is Fick's constant, ∇C is the concentration gradient in M m^{-1} and the orbital effect upon ions in the $\hat{\theta}$ direction is assumed negligible at macroscopic levels.

The retardation effect of the B field upon the diffusion rate may be estimated. For example, if B is 60 μT, a typical LGF strength, an estimate of relative difference in D $(\frac{\mu_{\mathrm{Ca}}^2 B^2}{24})$ is computed as 6.9 ×

[a] In fact, if the collision periods were all constant, and synchronous with the cyclotron frequency, there would be no transverse diffusion, because the ion would orbit one complete transverse circular path.

10^{-25}. Such a tiny difference due to the unenhanced LGF is likely inconsequential in terms of biological effect. However, in certain localised regions, this ratio may be increased. Biogenic crystalline magnetite chains (Frankel, 1986a) cause an enhancement of the LGF near chains of length a where the field depends upon negative powers of the distance r. Frankel used equations for a dipole appropriate where $r >> a$, and also for a chain of current cylinders, to show that the induced B- field, proportional to $1/r^3$, could be as high as 0.5 T, so the relative difference in D is 4.0×10^{-17}. It can be estimated that to achieve a 10% reduction in Fick's diffusion, a B field of strength 2.5×10^7 T is required. This estimate agrees well with other estimates (Kinouchi et al., 1988; Chiabrera and Bianco, 1987). In terms of the ionic flux and changes to it due to the B field, it is required to know the concentration gradients in the biological environment of interest.

Turning to the orbital effect, we estimate its rate from Eq. (4.15) for $B_0 = 1$ T. For a calcium ion, $\omega_c = 4.8 \times 10^6$ rad s^{-1}. For $t = 1.4 \times 10^{-14}$ s,

$$x = \frac{1}{2.4 \times 10^4} \sin\left(6.72 \times 10^{-8}\right),$$

$$y = \frac{1}{2.4 \times 10^4} \{1 - \cos\left(6.72 \times 10^{-8}\right)\}$$

Since $\omega_c t_{col} << 1$, we estimate the x and y offsets after one collision. Examination of Fig. 4.18b shows we may estimate $\Delta\theta$ as 3.07×10^{-7} rad per collision. Where the collision rate is 7.8×10^{13} collisions per second, the angular velocity of the orbit equals 2.4×10^6 rad s^{-1} $\approx \omega_c/2$. For a weak B field, $B_0 = 60$ μT and the angular velocity is 192 rad s^{-1} $\approx \omega_c/2$ again. The radius of the orbit is of the order of the mean free path length, about 5% of the atomic radius for a monatomic structure.

Depending on the mass, size and viscosity of the liquid, rotational mobility may induce biological effects on ions and dipolar proteins, including the disruption of the ability of tissues to attain and maintain homeostasis. Hence development may not be able to be initiated as normal while the nucleus or other biological elements are rotating. There is experimental evidence to show static ELF H fields and electromagnetic fields (EMFs) induce a range of bioeffects both harmful and beneficial. The window effects observed by Adey and others may be due, in part, to rotations where the alignment of cells is needed before ion flow occurs across the membrane.

4.11 Mechanisms behind Frequency and Amplitude Windows

Recapitulating for a moment, an ion of mass m and charge q moves in a static magnetic field of strength B with velocity v around a circle of radius$r$, where $\frac{mv^2}{r} = Bqv$, so $\omega = \frac{Bq}{m}$. This is the well-known case for ions moving in a cyclotron motion in free space. Moving in a liquid does not affect the process; there will still be magnetorotation as we have investigated. Whether it is significant biologically requires looking into the details of the exposure. Magnetorotation can occur in physical and biophysical situations. In physical situations there is much interest in the way black holes accrete by magnetorotational instability. This subject is outside the present text except to say that it is a highly numerical study that depends on ionised disks of matter that rotate as binaries: a star plus a black hole, or a neutron star, or a white dwarf, at the end of a star's life on the main sequence of its stellar evolution. This work can be described by the equations of magnetohydrodynamics. In the terms of SFT, we might call this a magnetoacoustic interaction. In the terminology of fluid dynamics, however, we talk about viscosity, which we have used in this Chapter to simulate ions in a liquid, and Reynold's number. In this cosmological environment, such rotations only occur under certain conditions; it is not a given that the disk will rotate to oblivion even if there is a magnetic field in place. What is needed are the precise conditions for these types of instabilities to form.

Something similar may be occurring in cellular biology. Obviously as with all science we need the precise conditions for magnetorotation to work. In normal terrestrial circumstances there is always a ubiquitous magnetic field present, the geomagnetic field. So if an ion were being magneto-rotated there could be a radial distance equal to the cell size. But we do not have an ion; we have a dipole. So what is the physical cause of the rotation? If we look at an elementary model of gravitation we require two elemental mass conglomerates. Now the motion will be a gravitating system like the sun and the earth, where each dipole will rotate about its dipole axis and both will move in an orbit, a self-field solution, a mutual dipole–dipole interaction. Finally if there is only one mass particle that is being exposed to an EMF this is an EM excitation of the dipolar mass and not a mutual interaction. Now more modes are possible. The single dipolar particle will spin about its charge axis and rotate in one of many radial modes with a circular orbit in several discrete orthogonal directions. This mechanism will be dependent on parameters, including EM frequency and EMF amplitude. So far we have not considered the biological cell. One of these modes may well have the correct direction and the correct circular radius as a best fit to a cell's actual shape, which may not be circular but which might still support the dipolar motions that have been excited by the EM source. And this is exactly what Adey discovered.

Reviewing Adey's window effects Postow and Swicord concluded:

> Numerous biological responses to electric and magnetic field exposure that involve fiequency or amplitude 'windows' have been reported. The majority of these responses have been observed when the biological system is exposed to ELF signals, ElF-modulated RF signals, or millimeter-wave signals. There is some consistency in the ELF and ELF-modulated. RF data; i.e., plausible arguments can be advanced to explain the similarity or differences of reported results. However, not enough data exist to be used as the basis of an explanation of the observed responses, or in some cases, firmly establish the nature of the response. There is no consistent repeatable pattern of millimeter-wave responses. Very few experiments have been repeated, duplicated, or confirmed by a second investigator. In some cases where duplication has been attempted, the original results were not confirmed. These experiments do, however, require a high degree of precision in the engineering aspects of experimental design, and lack of such precision could be a reason for non-reproducibility. For pulse-modulated effects, no consistent pattern of results has evolved (except for those effects involving a microwave-induced thermoelastic expansion of a biological tissue). This is most likely due to the limited experimental protocols that treat too few

experimental parameters and are dictated by equipment available to a particular investigator.

The existence of highly specific non-monotonic responses to electromagnetic radiation, at microwave and lower frequencies, is both unexpected and jarring to our sense of simplicity and order. This shock, and the appearance of a few controversial reports, have combined to produce a skepticism on the part of some observers of the field. However, skepticism can be constructive if it leads to the formulation of critical tests. At this time the body of experimental data is not sufficient to unambiguously verify existing theoretical models or even to indicate whether some of these models are useful guides to furlher experimentation. Often model-building efforts have not been detailed enough to predict decisive experimental outcomes. New measurement approaches and theoretical models are needed to improve our understanding of how electric and magnetic fields interact with living systems.

(*Source*: Postow E., and M.L. Swicord, Modulated fields and 'window' effects, in Polk C., and E. Postow (Eds.), *Handbook of Biological Effects of Electromagnetic Fields*, CRC Press, Florida, 1986.)

Adey's work was comprehensively reported in the last chapter of the review of Polk and Postow, and the conclusion, cited in full above, was the final comment in the book.[a] Yet it is *not* the final word on window effects. CEM and sciences based on CEM are currently incomplete. Specifically there is a lack of understanding at the photonic level. In the atom, for instance, SFT finds there are two spinors acting on each particle at the EM level, namely, the electron and the proton. The two EMFs act between centres of rotation and not between charges. Hence a proper understanding and appreciation of the macroscopic measurements has trailed behind the reality of the reactions occurring at the photonic level. It is little wonder that 'shock' and 'scepticism' have been found within the experts and the bioelectromagnetic (BEM) community. While the results were found, they could not be understood. They were 'shelved'. Now an understanding of the measurements is at hand,

[a] Polk and Postow wrote each occurrence of the term 'window' within the review in quotes as 'window'. Without understanding the basis of their scepticism, it was, in all likelihood their theory, CEM that was incorrect rather than the window experiments.

and they can no longer be forgotten, out of sight, out of mind. In ancient times till the 20th century, lepers were sometimes isolated on islands. Window effects of EM exposures have been similarly 'isolated' and need to be reviewed with a new knowledge of EM effects on dipoles. As we shall see in Chapter 5 a similar history of discovery, the failure to replicate and rejection occurred with Gurwitsch's 'mitogenetic' experiments in the 1920s–1930s.

A recent news item highlights need for ongoing, fresh research into health effects of power lines and EM appliance, given SFT's insights:

> Fetal exposure to magnetic fields from appliances, power lines may up kids' asthma risk—August 1, 2011 HealthDay News. A new study suggests that the children of mothers exposed to high levels of magnetic fields during pregnancy are at increased risk of developing asthma, findings that are sure to reignite the controversy over the health dangers that might be posed by exposure to power lines and electronics.
>
> Though the study does not establish cause-and-effect, researchers found a strong association between asthma in offspring and pregnant women's exposure to magnetic fields emanating from power lines and household items such as fluorescent lights, copy machines, electric blankets, microwaves and hair dryers.
>
> 'If EMFs [electromagnetic fields] truly increase risk as we have shown here, because of the ubiquitous exposure to EMFs the public health risk is serious,' said study author Dr. De-Kun Li, a reproductive and perinatal epidemiologist at the Kaiser Permanente Division of Research in Oakland, Calif.
>
> The study is published online Aug. 1 in the *Archives of Pediatrics and Adolescent Medicine*.
>
> Li and his colleagues followed the children of 626 pregnant women in Northern California for up to 13 years. During pregnancy, the women wore a meter for 24 hours that measured their average daily exposure to magnetic fields. Exposure was then divided into three groups: low, medium or high.
>
> Nearly 21 percent of the children, or 130, developed asthma.
>
> Children whose mothers had the highest magnetic field exposure (90th percentile or above) were 3.5 times more likely to have asthma than the kids of moms with the lowest exposure (10th percentile or less), the investigators found.

The children of mothers who were in the middle group for magnetic field exposure were 74 percent more likely to have asthma than kids of moms in the lowest group.

Put another way, about 13.6 percent of children whose mothers had the least exposure to magnetic fields had asthma, compared to 20.3 percent in the moderate group and 33.3 percent of the children of mothers with the highest magnetic field exposure.

'This is a carefully executed and analyzed study with a very provocative finding,' said Dr. Jonathan Samet, a professor in the department of preventive medicine and director of the Institute for Global Health at the University of Southern California, who has studied the health impacts of electromagnetic fields. 'This initial finding needs replication; the association is strong and merits follow-up.'

The study also found strong associations between EMFs and asthma in women whose children had other risk factors for developing asthma, including the mother having asthma herself or being the first-born child.

The children of mothers who had asthma and were exposed to high levels of EMF had six times the chances of having asthma, while first-born children had a 40 percent increased risk of having asthma if their mothers were exposed to high levels of EMFs.

Overall, about 13 percent of U.S. children have asthma, according to background information in the article, while prevalence has risen 74 percent from 1980 to 1996. The rapid rise has led experts to search for environmental exposures that might be contributing to the increase.

Magnetic fields are ubiquitous in modern life. Anything electronic, such as appliances and power lines, generate magnetic fields. Cell phones and other wireless technology are a source of EMF, although in the study, the device worn by the women was only able to measure lower frequency magnetic fields, such as those from household appliances and power sources. It was not able to track exposure to higher-frequency electromagnetic fields coming from wireless networks and cell phones.

Concern over EMFs was first raised decades ago. Prior research has searched for a link between EMFs and cancer. But most of those studies have found no association, leading to a 'dismissive attitude' about EMFs among many experts, Li said.

(*Source*: http://www.nlm.nih.gov/medlineplus/news/fullstory_114885.html)

Yet the experts are apparently still, somewhat ambiguously, unconvinced:

> The World Health Organisation issued Fact-sheet No. 263 in October 2001 on ELF (Extremely low frequency) EMFs and cancer. It said that they were 'possibly carcinogenic', based primarily on IARC's similar evaluation with respect to childhood leukemia. It also said that there was 'insufficient' data to draw any conclusions on other cancers.
>
> In May 2011, the WHO's International Agency for Research on Cancer published a review of the evidence on health risks of EMF, and concluded that there was limited evidence that cellphone users might be at increased risk of glioma and acoustic neuroma, and that there was inadequate evidence of any other health risks posed by EMF.

(*Source*: http://en.wikipedia.org/wiki/Electromagnetic_radiation_and_health)

In press release no. 208, also dated May 2011, the International Agency for Research on Cancer (IARC) classified RF EMFs as *possibly carcinogenic to humans.* To the experts there is 'limited evidence', 'inadequate evidence' and 'insufficient data'; there is confusion about the results. The window measurements are an 'aberration'. But this view also depends on an incomplete CEM theory.

In contrast, SFT provides a complete theory of EM. Adey's experiments are seen as a fascinating insight into EM excitation of dipoles constrained within cell membranes, a phenomenon supported by the mathematics of SFT. The excitation shown in Fig. 4.21a corresponds to the case of windows of ions moving across cell membranes. If the orbital radius is approximately equal to the mean radius of the cell then there will be a rotation and ionic currents will not flow across the membrane. If, on the other hand, the orbital radius is not equal to the cell's mean radius, then the dipole will not rotate significantly and transmission of ionic currents will be, in the main, unaffected. The dipoles in the cases shown in Fig. 4.21a,b

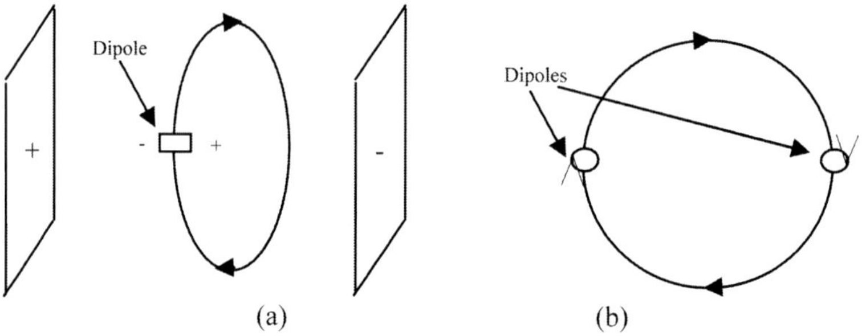

Figure 4.21 (a) Dipole[a] rotates exposed to EM by plate applicators, similar to Adey's window experiments (monopole–dipole); (b) dipoles gravitate about each other (dipole–dipole). Dipoles in cases (a) and (b) are in different orthogonal planes to the orbit; case (a) being a monopole–dipole interaction explains the orthogonal direction of the dipole compared to case (b); in both cases the orbit can be varied via the EM frequency or its magnitude.

are in orthogonal directions to each other, although both are still orthogonal to the direction of the orbital motion. The case shown in Fig. 4.21b is a mutual self-field interaction, similar to a solar system where a sun and its planets move around each other. In the case shown in Fig. 4.21a, however, the situation is excitation of a dipole by a 'monopole' EM source where other excitation modes and orientations are possible. We have only mentioned the rotational effect; in reality there is an additional dielectric effect involved in the window effects due to the alignment (and misalignment) of a colony of cells. In the analogous case to that shown in Fig. 4.21b of two single atoms interacting, for example, two hydrogen atoms, the forces are a dipole–dipole interaction and the result is a hydrogen molecule, which can be manipulated via frequency and amplitude of the exposure signal. With this SFT knowledge, a fresh research effort is needed to study the window effects.

In Chapter 5 we look at a mechanism where both the rotation and the alignment within a colony of cells are utilised to achieve division. It is likely that these effects are statistical in nature due

[a]The dipole-dipole talked about in Fig. 4.21 a-b is an effect between the differential of an E field and another differential of an E field. These dipoles are consistent with the concept of dipole in electrical engineering. The 'monopole' referred to in Adey's window experiments is an interaction between an electric-field and a differential E field.

to the specifics of the various exposure situations and the size and architecture of the range of human cells. Since the human cell is only approximately spherical it may be that any effects are dependent on the health and physiology of the individual involved in each case. Perhaps 'electrosensitive' and 'magnetosensitive' persons having unusual physiologies need to be considered as with individuals who have a resistance in the dry skin of their hands and feet, enabling them to touch bare power wires without experiencing life-threatening electrocution.

In addition there are other phenomena that are supported by Adey's experimental measurements, the evolution of deoxyribonucleic acid (DNA) and later the evolution of the cell. We asked in Chapter 3 how these biological elements could have evolved. Was it the sun that provided the EM energy that rotated molecular combinations into forming amino acids with their spinor-like shapes and repeated molecular sections? Cells could then have used protein magnetorotation to supply the necessary levels of energy in order to achieve mitosis as cosmological expansion proceeded. Did the sun's excitation eventually evolve into a mutual interaction between dipoles in the membranes of cells that used their mutual effect to magnify currents within the DNA to assist mitosis?

We in the modern world, however, have pressing questions. Amongst other issues, what is the cause behind the rate of infertility across the technologically modern world where there is a marked difference between the developed and the undeveloped worlds? Could EM appliances be the cause of the modern levels of endemic infertility? On the other hand, if Adey's measurements are correct, what does this say about how the cell achieves mitosis? How does *frequency medicine* work at a mechanistic level? There both are hazards and opportunities; medical science must walk across the chasm between ignorance and knowledge.

To recapitulate, SFT gives fresh physical and biophysical insights:

(1) A dipole–dipole interaction by which the sun may have provided EM energy to kick-start biological evolution on the earth. This may have first been via the structure of amino acids, then the DNA molecule and finally cell structures with their plasma membranes, dipolar proteins and spindle poles that were eventually used to achieve mitosis.

(2) Support for the window experiments of Adey and others whose efforts were ultimately rejected by BEM and medical experts some decades ago on the basis of classical EM theory.
(3) A method to manipulate cells for medical benefit or therapy.
(4) A mechanism by which rotation of the cell or its nucleus, as in the example of Kirson et al. discussed in Chapter 1, can occur; as well as being beneficial, this can be deleterious to mechanisms of repair and maintenance within the body.
(5) A complete theory of EM giving a new mathematical description of physics and biophysics. This involves different forms of gravitation across the universe, including a new gravitational form at the galactic level, and solutions for the strong and weak nuclear forces in the universe. The cellular dynamics discussed above in the context of window effects are closely related to the forms of gravitation and their closed-form solutions given by SFT. Frequency and amplitude effects are seen in gravitational coupling between suns and planets, as well as in window effects observed in EM exposure of cells.

Chapter 5

Cell Division and Membrane Diffusion

Multi-cellular life forms require many cellular divisions within colonies to achieve growth and development. According to self-field theory (SFT) there are biophotonic stages involved in the overall process termed the cell cycle. While microbiology is largely based on atomic chemistry, still to embrace biophotonics, this situation is changing within mainstream academia. Cells occur in large populations and interact via cell–cell signalling of electric (E) and magnetic (H) fields. Chromosomes emit biophotons of specific frequencies within the IR and UV ranges. The divisions appear to be an SFT process where biogenic E and H fields arise from diffusion of dipolar proteins within the plasma membranes of a cell colony. The biogenic mechanisms of cell division include diffusional processes within metaphase and anaphase. Dipolar proteins diffuse within cell membranes in response to intra- or extracellular endogenous or exogenous E and H fields. These processes involve unconstrained protein diffusion and two feedback loops, (1) electrical feedback between membrane proteins of a fertilised cell and those of the colony and (2) magnetic feedback of biophotons emitted by chromosomes in the fertilised cell in the colony. Gurwitsch's observations of cell divisions of the 1920s–1930s, and α-region dielectrics at ELF frequencies support the SFT

Inside the Photon: A Journey to Health
Tony Fleming

ISBN 978-981-4241-40-3 (Hardcover), 978-981-4241-88-5 (eBook)
www.panstanford.com

hypothesis. Now, biophotonic research of growth and development has a field theory to test against the biology.

5.1 Introduction

In *The Origin of Species by Means of Natural Selection, or the Preservation of Favoured Races in the Struggle for Life,* Darwin presented his work in 1858 to the Linnaean Society, causing an explosion within the halls of science. Proving the pen is mightier than the sword, Darwin created no less of a revolution *outside* the scientific domain. Antagonists were left reeling. Darwin's discussion of embryos thrust embryologists and histologists into the spotlight at the turn of the 20th century (see Fig. 5.1a,b). Today epigenesis and preformation are opposite views to explain the development of the individual form. Science sees the recent debate between the two camps in a wider perspective, going back to Aristotle and his interest in chicken embryos. In this view 20th-century genetics represents a new form of pre-determinism, while the 21st century may be a new era of epigenesis, where the cosmos itself is seen as the cradle of life. Stem cell research reveals the fate of cells is not gene specific. Each stem cell can be harvested and cultured into a number of adult cell types. Differentiation is normally specific

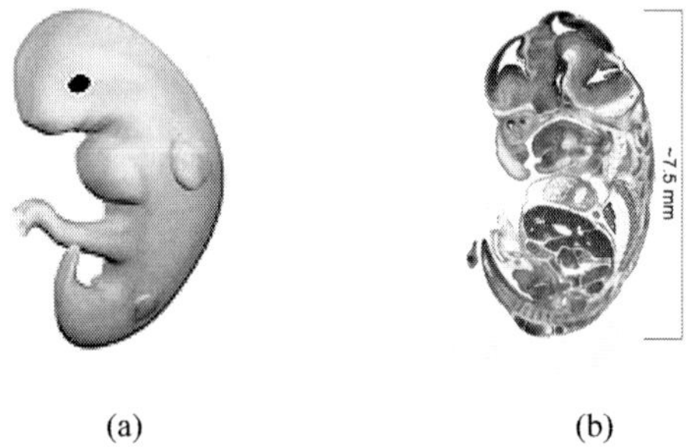

Figure 5.1 (a) Human embryo after six weeks and (b) histological film of mouse embryo after 10 days (credit: Wikimedia). Resemblance of embryos led Darwin to state that a 'community of embryonic structure reveals community of descent.'

in its direction; it was not suspected that other types could be obtained depending on the position where growth occurs within the organism. The debate continues, chicken or egg, nurture or nature, internal or external process. Microbiology and astrobiology supply some answers. According to SFT, at biophotonic and cosmological domains, life is a resonance phenomenon. Individuation depends on nurture *and* nature and has been an ongoing evolutionary process since the proto-earth emerged from the swirling gases of the early solar system. Biological evolution relates to the cosmological dynamics back to the big bang. Each human embryo and evolved being may stem from the universal 'embryo', the exploding kernel of the big bang when the fields were only a roaring sea of sub-photonic particles that reached us as the cosmic microwave background.

We have seen that biological evolution has cycles measurable over galactic time periods. SFT suggests that resonance of fields over many domains has been involved in evolution. The evidence also shows that biological evolution has had periods when, seen from the terrestrial domain, survival of the fittest was a driver behind change. At the same time cooperation between cells must have always been present for mitosis, cell division, to work. This cooperation was required to achieve the necessary energy density within the cell after the cosmological expansion linked to the big bang reduced the energy density. This strange mix of factors is all part of biological evolution. We find complementary factors involved in evolution. This is similar as to whether the photon is a particle *or* a wave. We find with SFT the photon is a particle *and* a wave. So too with evolution, rather than just survival of the fittest, cooperation between life forms was also a factor. When we come to the human cell we may ask, Is this a chemical- or a field-based entity?

A lack of emotional control is observed in some murderers. The relationship between frontal cortex damage and homicide gives insights into our emotional makeup. Functional magnetic resonance imaging reveals a lowered ability to process and mask anger in such damaged offenders. At the start of the 21st century, a mix of internal and external factors is seen to be involved in our emotional fabric, just as our deoxyribonucleic acid (DNA) changes from generation to generation. By understanding how cells respond to fields we may learn to regrow such tissues previously damaged. We know of tissue engineering, gene therapy and genetically modified foods. Can we modify cells via biophotonics, by imitating how the biogenic electric (E) and magnetic (H) fields work? Can we use electromagnetic (EM) fields in a more refined manner of therapy than pharmaceuticals, genetics or macroscopic engineering alone?

The genetic information stored within DNA is an internal form of bioinformation that is unfolded and unpacked, as needed. Each page is read and the chemical instructions carried out. As each page of coded instructions is implemented, the outside world inputs via molecular and biophotonic means. Incubation requires an environment that is supportive of the overall process, not just biochemical particulates, but a nurturing medium, including the fields and the ambient energy bath. In terms of an analogy, we may think of DNA as consisting of letters and words forming books with pages. The various proteins are the managers who read the words on the book's pages and delegate the work. Yet the various fields do the actual work of growth and development.

Knowledge of chemistry has been limited by the inherent inaccuracy of quantum mechanics (QM) with its Heisenberg uncertainty principle. Molecular chemistry, the nanoscopic view of biology, is incomplete. SFT sees gravitation all the way up to and including the cosmological level as a form of molecular interaction. Till recently this lack was not understood, even though Einstein flagged the problem before World War II (WWII). Our health and safety standards are still at present written around the macroscopic tenet of temperature changes to a piece of tissue one cubic centimetre in volume. Classical EM, as given by Maxwell's equations, is linked to present-day macroscopic bioeffects. This is not a sufficiently detailed view to understand embryology. We need to understand the biophotonic level of interaction. How do cells replicate via the fields that are biogenic to tissues? And where do the fields come from? By understanding these field mechanisms and how they percolate into action, we progress. This is achieved via SFT that sees the overall division process as a series of balances followed by transitions and eventually cascades across molecular states. It is not only via atomic-level chemical processes but by the interplay between fields and biomolecular structures that the processes involved in division occur. As with the question whether the photon is a wave or a particle, the question is whether the cell is a metabolic *or* a field entity; the answer is *both*.

5.2 Understanding Cellular Division

The self-view of science at the start of the 20th century included an eclectic hierarchy of disciplines ranging from astronomy, palaeontology, geology, physics, ponderable matter (interstellar space; at that time the ether was prominent in scientific debates), meteorology, physics, chemistry, anatomy and physiology, chemistry, biology, medicine and psychology. Not unnaturally, geology and palaeontology associated with Darwin's theory of evolution featured near the top of this list. The entire world, including science, wanted to be associated with Darwin's evolution and its survival of the fittest. Microscopy had a profound effect on this early history. Light can only be used down to micron-sized entities; hence cell interiors were difficult at best to study. The nucleus could just be seen as a dark spot at the centre of the cell. This fundamentally limited the study of microbiology till 1931, when Ruska and Knoll invented the electron microscope. The main experimental thrust at this time was embryology. By cutting and pasting sections of embryos and transplanting to other than their normal position, new or multiple limbs and organs could be grown in these alternative positions. Like art imitating life, the story of Frankenstein came from Mary Shelley's mind some decades after Galvani's revolutionary work with bioelectricity and the twitching of dead frogs' legs. In many ways Frankenstein was a morality tale about the dangers of unfettered science. In the age of eugenics in the early 20th century, science began cutting and pasting to investigate the embryonic effects.

In this environment, evolutionary biology acted within the revolution in scientific thought needed to accommodate biological evolution and life at the scale of the micron. Science underwent intense specialisation in the 20th century in which many areas of biology originated. One of those areas was microbiology, defined in terms of 'small life'—microorganisms such as bacteria and viruses (Fig. 5.2). It is the domain of life at the micron in the same way

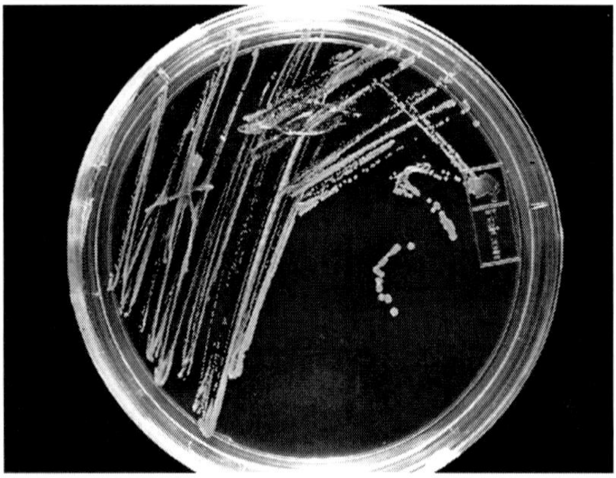

Figure 5.2 Agar plate with colonies (credit: Wikipedia).

that nanobiology is the domain of life processes at the nanometre, molecular biology. The trouble was that 20th-century science could not see the forest for the trees—too many islands of fact and not enough interconnections. The problem was that the connection was the biophoton, a tiny quantum of energy misunderstood by microbiologists, biophysicists and biochemists alike, except a handful.

It is known by few today that in the early 20th century theories arose as to how fields were involved in the mysterious growth seen under the microscope. Harrison in 1907 introduced a new technique, *tissue culture*, for studying growth, at the same time solving the origin of nerve fibres. Gurwitsch in the 1920s observed ultraviolet (UV) radiation being emitted by living systems, including onions and yeast. As shown in Fig. 5.3 he placed two onion roots in close proximity to each other. One onion root was growing, and this influenced the growth at the tip of the other root. Although photon counting was not possible at this time, he placed window glass and ordinary glass between the roots and was thus able to determine the light was UV in nature. The work suggested a link between UV photon emission and cell division. Gurwitsch called this 'mitogenetic' radiation. An era of intense interest grew until just before WWII. Unfortunately, like Adey's window effects of the 1970s, some failed to replicate Gurwitsch's mitogenetic

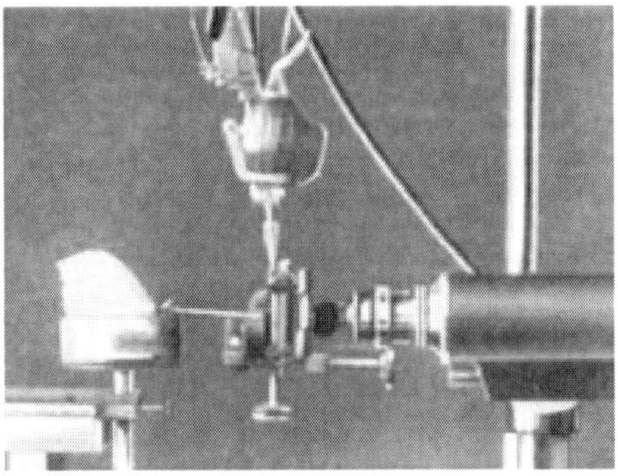

Figure 5.3 Gurwitsch's onion root experiment (*source*: http://www.21stcenturysciencetech.com/articles/summ01/Biophysics/Biophysics.html).

experiments. Meanwhile, biology moved on to a golden age of genetics supported by the numerical calculations of QM and the discovery of the helical structure of DNA. The molecular viewpoint now dominated, and the field view remained unsupported for several decades until Quickenden and Popp from the 1970s onwards used photomultiplier techniques to successfully determine UV fields were truly involved in cellular dynamics. These techniques originated in astronomy where very weak photon streams could be accurately measured across the universe. Using modern photon counters the biophotons from strands of DNA were found to be coherent, as if the DNA were acting like a biological laser.

In general, from around 1970, the microbiological community no longer saw UV radiation as centrally involved in division and growth. Rather it was seen as a by-product, a result of what was considered to be a more fundamental biochemical action. Yet once realised, the field view did not go away. In 1996 Fleming saw the role of membrane protein diffusion in attaining the amazingly high dielectric permittivities of some tissues such as the liver ($\varepsilon_r^{\text{liver}} \approx 10^7$). In this view the membrane proteins of many neighbouring cells polarise and align, causing enormous magnification of the original signal from a single cell. Measurements reveal this amplification occurs for both static and time-varying E field sources. A corresponding rotation of the membrane proteins occurs with both static and time-varying H field sources, perhaps when the particular biological tissue is not especially magnetic, a diamagnetic effect. Hence we are looking not only at a single cell but also at a population of neighbouring cells cooperating to cause enormous magnifications of the originally tiny biogenic signals. Chromosomes near metaphase are central to the effect. Many cells may be involved, continuing the process of cellular division into tissue growth. Many processes of growth, maintenance and repair may use such fundamental cell–cell dynamics.

At the start of the 21st century new areas of specialisation within biological growth and development, such as stem cell technology and tissue engineering, promise replacement or regrowth of diseased or missing organs and tissues. This surgical approach, like microbiology in general, is devoid of substantial knowledge of the role of the biogenic E and H fields at the microbiological level. While these technologies will likely become life-saving, organ-regenerating tools for people across the globe with a range of diseases, a therapeutic procedure that does not involves surgery or pharmaceuticals must surely be a desired objective in healing. Biophotonic therapies may in the future be able to be applied from the cradle to the grave without fear of side effects, being applied as forms of therapy. While surgery and pharmaceuticals have been indispensable as we have lived through the 20th century, they may in time be seen as options in the overall armamentarium at the call of the overall medical system as this present century proceeds. The scientific method needs to be applied to the emerging field theory of cell dynamics. We now discuss this SFT hypothesis on the basis of the interactions between the chromosomes and the motions of dipolar membrane proteins within a colony of cells.

5.3 Protein Diffusion within the Plasma Membrane

Developmental microbiologists think of cell–cell cooperation as chemical signals carried via proteins. The past 70 years have taught that atomic chemistry is the main tool for analysing the cell. The powerful methods of genetics have meant that an important part of the actual mechanics of signalling and organising between cells, the E and H fields, is overlooked in the current chemical paradigm of

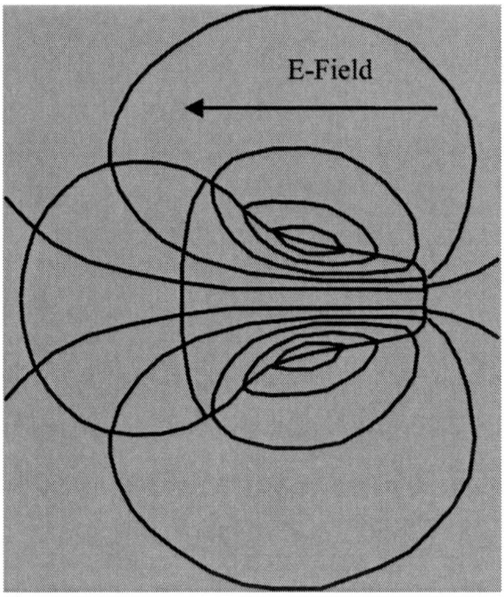

Figure 5.4 Growth of a cell tip.

how cells work. In the following we look at how cells use membrane proteins to act cooperatively to achieve some of the basic steps of division. Mitosis, for instance, contains many steps, including stages of the cell cycle. Genetic signals are responsible for these stages that are kept within the complete genome, which for *Homo sapiens* is 3.4×10^9 base pairs (bp) long. However, E and H fields play an important role in adjunct (Fig. 5.4).

A numerical study of protein diffusion within the plasma membrane of a cell exposed to a static E field demonstrates the mechanism for a single cell. A finite-difference method was used to mathematically model the equations of membrane diffusion proposed by Jaffe and Nuccitelli (1977) and Poo (1981). A computer program was developed and used to compare theory with experiment. Several runs were performed varying the cell radius (10–100 μm), exposure level (100–1,000 V m^{-1}) and various other parameters, including the diffusion constant. Next, the diffusion constant was matched to the time to reach electrophoretic equilibrium observed by Poo after exposure of *Xenopus* myotomal cells. The numerical results compare reasonably well with experimental time-sequence asymmetry data for concanavalin receptors reported

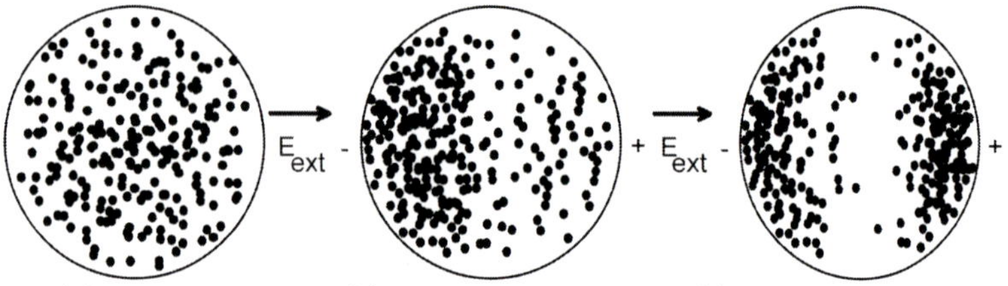

Figure 5.5 Diffusion of proteins within the membrane surface (a) before an external E field is applied, (b) after an external E field is applied and there is a net negative charge on the surface and (c) after an external E field is applied where a dipolar charge is induced within proteins.

by Poo. A final run was made at weak exposure levels to model the experimental exposure of Mougeotia protoplasts conducted by White et al. (1990).

Electric fields are known to cause proteins to diffuse within plasma membranes. Darnell et al. (1986) discuss freeze fracture micrographs that clearly show diffusion of intracellular proteins within the plasma membrane of mitochondria due to applied voltage gradients resulting from strong E field exposure. Figure 5.5a–c illustrates how asymmetric and symmetric protein concentrations develop across a cell as diffusion proceeds in time. Jaffe and Nuccitelli (1977) and later Poo (1981) investigated protein diffusion by means of a chamber to which immobilised cells adhere, enabling in situ electrophoresis of membrane components. Agar-filled salt bridges are used to pass current through the saline-filled chamber. Fluorescent labelling, micro-iontophoresis of membrane components involved in ion transport across the membrane and the freeze fracture method are used to experimentally determine the migration of proteins and other components. Exposure levels in these experiments were between 100 and 1000 V m^{-1} (Poo, 1981).

Protein diffusion within membranes is involved in the growth and development of cells via processes that are initiated and controlled by biogenic E fields. In one such investigation, spherical *Mougeotia* protoplasts, originally cylindrical, were exposed to static E fields of around 20 V m^{-1} (White et al., 1990). The protoplasts regenerated, displaying a normal sequence of microtubule reorganisation that aligned perpendicular to the applied field. The plasma membrane in the region attached to the microtubules was now observed to stretch in the direction of the applied field and re-establish the original cylindrical shape. How regrowth occurs is unclear; one possibility concerns the electrophoresis of membrane components (Poo, 1981). Mechanical stress due to the applied field could be the cause of stretching (Bryant and Wolfe, 1987).

In another study, effects of exposure to static E fields of 1,000 V m^{-1} upon the cytoskeleton were noted in experiments designed to observe and quantify galvanotaxis of fibroblasts (Hui, 1994). Changes in intracellular calcium flux were measured using aequorin, a calcium chemiluminescent indicator protein, and fluo-3, a calcium-sensitive fluorescent dye. The motility of the fibroblasts was measured, and fluorescent micrographs showed changes in actin-containing stress fibre patterns of the cytoskeleton that span across the cell body and are attached via proteins to the inner plasma membrane. Such changes to cytoskeletal structure are difficult to quantify, requiring sensitive methods of measurement.

While some obvious effects of protein diffusion are indisputable (Darnell 1988), the bioeffects community regards developmental effects due to weak exposure levels and cytoskeletal effects that are difficult to quantify as yet to be understood and thus still in question (Carpenter and Ayrapetyan, 1994). There is a need then, in parallel with experimental techniques, for numerical efforts to obtain quantitative understanding of cell growth and development, especially where longer exposure times and weaker levels of exposure are involved.

As there is probably an underlying combination of causes (Carpenter and Ayrapetyan, 1994), the weak effects discussed above may depend only, in part, upon diffusion of proteins within the plasma membrane. Quantification of protein diffusion requires precise knowledge of the components involved in the process, and further efforts are required to understand the process of protein diffusion itself. One factor requiring clarification is illustrated in Fig. 5.5b,c. Particles diffuse differently if there exists a permanent net charge on them or whether they are capable of being electrically polarised. The range of electrophoretic particles requires that their electrical nature, diffusion and mobility characteristics be determined before accurate diffusion models can be obtained. It is possible protein diffusion is a combination of both types of diffusion. Our aim, however, is to model protein diffusion using current knowledge, which is based on the asymmetric process shown in Fig. 5.2b.

5.4 Theory of Protein Diffusion

Below is an outline of current protein diffusion theory (Jaffe and Nuccitelli, 1977; Poo, 1981):

(1) Biogenic E fields that exist in the extracellular medium will in general induce fields that are both normal and tangential to the membrane. There will be tangential components of the induced external E field at all points on the outer membrane surface. These tangential E field components vary from point to point on the membrane surface, depending on the response of the cell to the external E field. Solving a field problem involving the cell and the exposure set-up, we can obtain the tangential components. In the theory presented by Jaffe and Nuccitelli (1977) and Poo (1981), the membrane is considered non-conductive, making the internal fields zero. The E fields just outside the membrane have a simple analytic form

$$\vec{E}^{tan}(\theta, \phi) = 1.5\vec{E}_{ext} \sin\theta \tag{5.1}$$

where $\vec{E}^{tan}(\theta, \phi)$ is the tangential field just outside the membrane, $\vec{E}_{ext}(\theta, \phi)$ is the external source field and θ is the elevation angle and ϕ the azimuthal angle (see Fig. 5.6). We assume the cell to be slightly conducting, and hence E fields exist inside the cell in response to the external field. While Eq. (5.1) is a good approximation since the membrane conductivity is relatively low, the actual induced E fields will be slightly lower than as given by Eq. (5.1) due to some of the external field being transmitted into the cell.

(2) Due to these extracellular tangential E field components, the proteins can diffuse within the plasma membrane according to normal diffusive principles. Taking a surface diffusion process

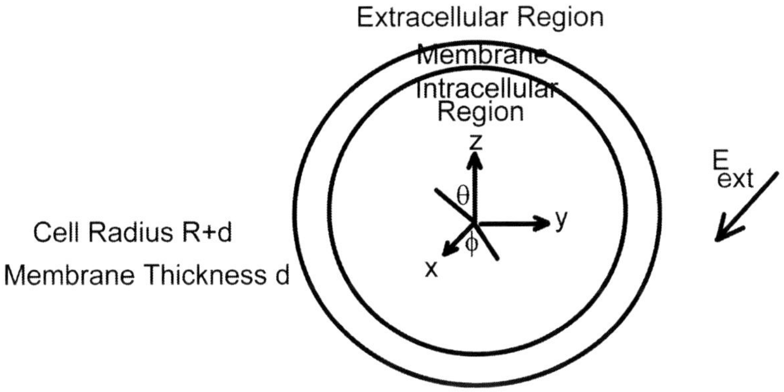

Figure 5.6 External E field along the x axis, incident upon a two-layered spherical model of a cell.

instead of a volume process, we write a continuum equation for protein flux:

$$j_{\mathrm{P}} = -D_{\mathrm{P}}\nabla C + C\mu_{\mathrm{P}}\vec{\mathrm{E}}^{\tan}(\theta, \phi) \tag{5.2}$$

where j_{P} is the protein flux in particles in $\mathrm{m}^{-1}\mathrm{s}^{-1}$, D_{P} the protein thermal diffusion constant in $\mathrm{m}^{-1}\,\mathrm{s}^{-1}$, μ_{P} the protein mobility in $\mathrm{m\,s^{-1}/V\,m^{-1}}$, C the concentration of proteins in particles in m^{-2}, ∇C is the protein concentration gradient in particles in m^{-3} and Eq. (5.2) is known as the Nernst–Planck equation[a] and can be used to estimate the migration in time of surface proteins within the membrane due to thermal diffusion and any E field present.

(3) There is a net negative charge on the outside surface of most cell membranes. Hence an externally applied E field will induce the negative charge carriers to the cathodal (-ve) side.

(4) The diffusion process is assumed to result from the net surface charge, and this leads, after electrophoresis, to an asymmetry in the concentration of proteins at the cathodal and anodal

[a]A similar equation for ionic flux across the membrane exists in parallel to Eq. (5.2), thus forming a 'double diffusion' process. An imbalance anywhere across the cell membrane between ionic flux and transmembrane potential may cause dipolar proteins to be polarised by the induced component of the E field normal to the membrane. This may form another protein diffusion mechanism, apart from the suspected net negative charge process presented here.

sides, as illustrated in Fig. 5.2b. This asymmetry can be used to estimate the ratio of the diffusion constant D and the mobility μ so that

$$\frac{D}{\mu} \approx \frac{3E_{\text{ext}}R}{\ln[(1+A_s)/(1-A_s)} \tag{5.3}$$

where the asymmetry index $A(t) \equiv [C(180^\circ, t) - C(0^\circ, t)]/[C(180^\circ, t) + C(0^\circ, t)]$ and $C(180^\circ, t)$ and $C(0^\circ, t)$ are surface concentrations of proteins at the cathodal and anodal sides at time t.

(5) The diffusion constant D can be related to the cell radius $(R+d)$, where R is the intracellular radius, d the membrane thickness and characteristic time for approach of equilibrium τ_e.

$$D = \frac{(R+d)^2}{\tau_e} \tag{5.4}$$

The transmembrane potential (around 100 mV) is directed normally across the membrane bilayer (radially in the case of a spherical cell). Its main purpose is to maintain a concentration gradient of ions across the membrane. If there were any tangential transmembrane potential we might expect that a non-uniformally shaped cell would result. For protoplasts, this is observably not the case, and reasonably rounded shapes are evident.

If dynamic equilibrium does occur, protein flux within the membrane due to thermal diffusion is balanced by the flux due to the external E field, E_{ext}, at which point the protein concentrations within the membrane do not vary as time proceeds (Plonsey and Barr, 1988). After the proteins have diffused in accordance with the electrophoretic forces set up by the external E field, other processes may now follow. For example, if there are clusters of proteins near one another, there may be enough interprotein forces to cause alignment into a regular array or chain, similar to ion–ion or dipole–dipole forces that occur in crystal lattices. Again, if some of these proteins contain ion-specific channels for the transduction of ions across the plasma membrane, an increase in ionic currents entering the cell occurs.

5.5 Finite Difference–Time Domain Equation of Protein Diffusion

A numerical model of protein diffusion was developed. The method used is a simple form of the finite difference–time domain technique that has been used to model diffusion across cell membranes (Sala and Hernández-Cruz, 1990). The method is simple to apply since numerical time derivatives are not needed, only numerical concentration gradients. Figure 5.3 illustrates the assumed geometry of a spherical cell having radius $R + d$ and membrane thickness d, exposed in the X direction to an E field of strength E_{ext}. As shown in Fig. 5.7, the membrane is divided into finite surface regions in spherical coordinates ($r_0 = R + d/2$, θ, ϕ where r_0 is constant. Grid points are assumed to exist at the centre of each sub-area. The network of grids is time-stepped using mobility, gradient and concentration information at the previous time step. Equation (5.5) represents an approximation to the actual continuously time-varying flux at a grid point on the cell surface at an instant in time by a simple approximate flux that is calculated at each grid point. Using distances between grid points in both θ and ϕ directions, an approximation for the number of proteins that flow between grid points can be calculated both for flux due to mobility and back diffusion, Eq. (5.2).

$$j(t = i\Delta t, \theta = j\Delta\theta, \phi = k\Delta\phi) \approx \tilde{j}_{i,j,k} \tag{5.5}$$

$$\tilde{j}_{i,j,k} = -D\nabla\tilde{C}_{i-1,\theta_j,\phi_k} + \tilde{C}_{i-1,\theta_j,\phi_k}\mu\vec{\mathrm{E}}^{\tan}_{\theta_j,\phi_k} \tag{5.6}$$

Corresponding to the indices in Eq. (5.6), the domain of the model was subdivided into n_{t} time steps, n_θ elevational sub-divisions and n_ϕ azimuthal sub-divisions.

Boundary conditions were applied at the north pole by allowing fluxes to flow over the top into the opposite azimuthal area. Symmetry was assumed for the southern hemisphere, so no flux flowed between the upper and lower areas across the equator. Calculations were needed only for the northern hemisphere. Areas connected by azimuth at $\phi = 0^\circ$ were assumed to be continuous with areas at the end ($\phi = 360^\circ$) of the azimuthal range.

Initial conditions were applied to the model by assuming an initial steady-state equilibrium. Protein populations within each

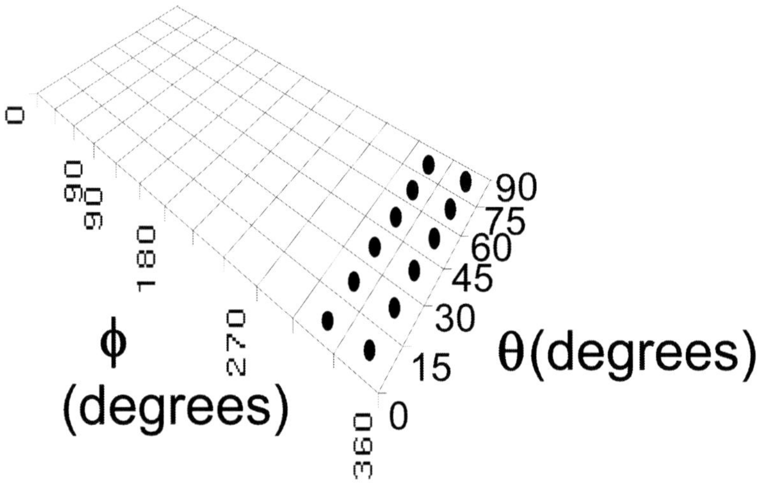

Figure 5.7 Membrane surface divided into finite areas showing two rows of grid points.

finite area of the grid network were determined by calculating each sub-area and assuming a total protein population. Although the total protein population was not known, a value of 5×10^9 was assumed using data from Darnell (1988). Hence the final results were in terms of assumed concentrations. The asymmetry index is unaffected, however, since it is a ratio of concentrations.

The E field problem to determine the tangential fields just outside the membrane surface in response to an external field was solved using the sub-division of the membrane surface, as shown in Fig. 5.7. This was done using a stand-alone code DSPH3L (Fleming 1996).

5.6 Response of the Cell to an External E Field

The response of cells to electrical stimuli is a complex process that is not yet completely understood either biologically or electrically. Jaffe observed ionic currents near the growth tip region of the seaweed *Fucus* (1979), demonstrating the existence of biogenic E fields (Fig. 5.8a).

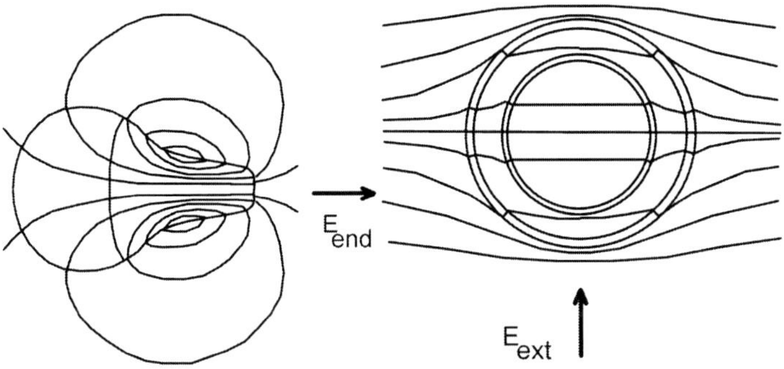

Figure 5.8 Symmetric and asymmetric growth. (a) Seaweed *Fucus* and biogenic ionic current in and around the growth tip and (b) lines of equal potential through a layered spherical cell due to an external E field.

It is known (Furnell, 1991) that in uniform fields, objects such as layered spheres and ellipsoids respond to external E fields by setting up equipotential lines, as shown in Fig. 5.8b. For biological cells, whose membranes act to shield the internal regions, these equipotential lines flow mainly around the cell; any that penetrate the first membrane will flow mainly around the next membraned region, etc., until the innermost region is penetrated. The small but finite conductivity of the membranes allows penetration by the external E field. Any lines in the inner region are analytically constrained to be straight. Lines of potential within the membrane regions tend to be radially directed with small tangential components. Pethig discussed the perturbing effect of small 'conducting pores' within the membrane (1988).

Fleming studied the complete EM and quasistatic formulations for layered spherical models of cells (1996). A cell of radius 10 μm was assumed to be exposed to a uniform external E field of strength 1 V m^{-1}, and located in a physiological saline solution (0.9% Na). The suspension of spherical cells and surrounding saline solution was assumed to be infinitely dilute to simplify the associated problem of estimating the dielectric constants for the computations. The simple two-layered geometry was as shown in Fig. 5.7. Using a similar analysis, a cell of radius 20 μm was examined, corresponding to the average of the cell sizes used by White et al. (1990).

Results are shown in Fig. 5.9a–c. In Fig. 5.9a, the total magnitude of the E field at the membrane mid-point ($r = R + d/2$) is shown across the upper-half membrane surface induced by a normalised external E field of strength 1 V m^{-1}. Corresponding to this result, Fig. 5.9b shows the induced θ component at these mid-membrane points, while Fig. 5.9c shows the same induced θ component just outside the membrane in the extracellular region.

Examination of the results shows that the radial component is maximal, just under 6,000 V m^{-1} (per unit incident field), at the leading edge on the mid-plane and at the opposite point of this plane on the other side, while going to zero at both poles. Correspondingly, the θ and ϕ components (ϕ not shown) act in opposite fashion, reaching a maximum at the poles and zero around the equatorial plane. Their total magnitude, however, just outside the membrane in the extracellular region, is reduced from the 6,000 V m^{-1} of the radial component, being, as expected, at the maximum, about 1.5 times the incident field. Interestingly, the tangential field inside the membrane is about 0.5 times the extracellular tangential field.

At this point, a word or two is required to discuss the size of the E field strengths we have just presented and how these relate to the endogenous membrane E fields, which are huge by comparison. First, assuming two elementary electric charges to be separated by a distance of 5 nm, the corresponding Coulomb E field is calculated to be 5.8×10^7 V m^{-1}. The potential across this distance is then 287 mV, which is of similar order to measured transmembrane potentials of around 80–100 mV (Plonsey and Barr, 1988). It can readily be seen that the huge E fields across a lipid bilayer tend to dominate the structure in which many lipids are located next to each other. Having no tangential components, the result is a regular lattice of very strong opposing lipids. The lattice exists to maintain a structural interface between intracellular and extracellular regions and also creates an environment in which the surface can maintain a concentration gradient of ions diffusing across it.

Do the small tangential components have any physical significance? We answer this by analogy. Examining the atomic structure of an ion, we see that similar, even larger, Coulomb forces are used

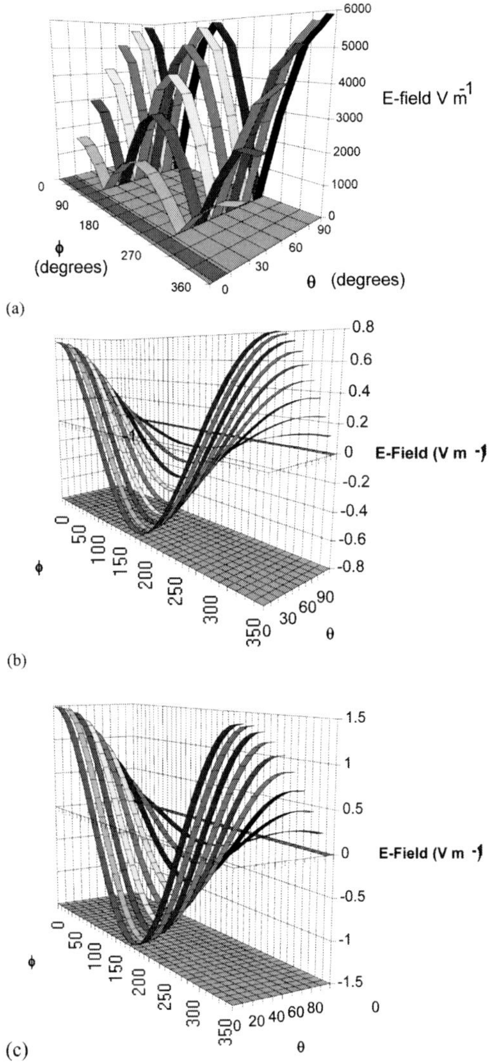

Figure 5.9 (a) Total E field (magnitude) on upper-half membrane surface $0^\circ < \theta < 90^\circ$, $0^\circ < \phi < 360^\circ$ in response to an external E field of 1 V m^{-1}. (b) θ component of an E field at the membrane mid-point ($r = R + d/2$) across the upper-half membrane. (c) θ component of an E field just outside the cell across the upper-half membrane.

to maintain structure—in this case electrons in orbit around its nucleus. These huge internal fields do not stop us assuming a very simple ionic model for analysing ionic motion, which we know as ionic currents and where we use E fields that are tiny, miniscule, by comparison. What is important is the net field that is present in the external world.

It also needs to be kept in mind that exposure times for some cellular effects involve quite lengthy periods. White et al. exposed protoplasts for around three to four hours before the appearance of two foci at the poles. Stochastic fluctuations, if significant, exist over very brief periods and can be expected to average out over long periods.

5.7 Electrophoresis of Proteins within the Cell Membrane

Using the θ and ϕ tangential E field results, which were stored in a file, the diffusion model was tested. Initially, a population check was made. This ensured that the same number of proteins were present at the end of a run as at the start. The check was found accurate to within 10^{-5} particles out of 5×10^9. The mobility and back-diffusion fluxes were programmed separately and the output checked for physicality. The results were symmetric in ϕ and showed an increasing effect as θ increased. Equilibrium was observed only when back diffusion was included.

Satisfied that the program was fully working, the complete program was tested using the results obtained by Poo (1981). In Poo's experiments, the asymmetry induced upon concanavalin receptors was tracked in time and exposure levels were varied. To obtain a comparison the diffusion constant used in the numerical model was obtained using Eq. (5.4) by assuming a range of cell radii and assuming an equilibrium time of around 600 seconds. Figure 5.10 shows the time-varying results for the case of a cell radius of 100 μm. The variation across a range of elevation angles, corresponding to the north pole down to the equator, is shown. The results show that asymmetry is foremost at the leading edge ($\theta = 90^\circ$, $\phi = 0^\circ$) but is prominent right along the leading azimuth ($0^\circ < \theta$, 180°, $\phi = 0^\circ$).

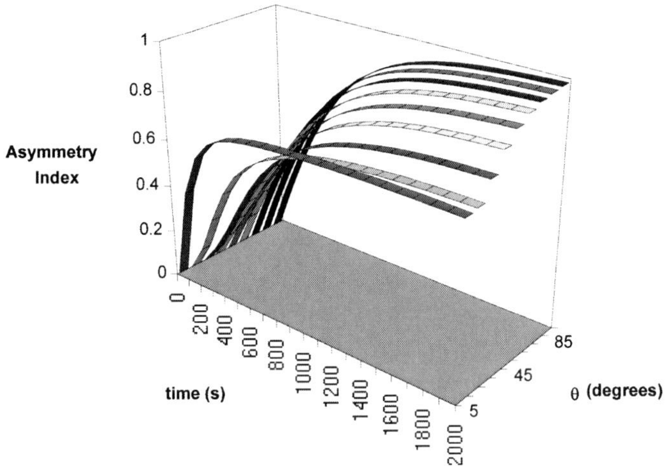

Figure 5.10 Asymmetry index vs. time for a cell of radius 10 μm over a range $0^\circ < \theta < 90^\circ$.

The asymmetry index is shown for cells of varying radii (10–100 μm) in Fig. 5.11. The results show that for a cell of radius 100 μm, equilibrium is reached in about 1,200 seconds. The other cell radii have not reached equilibrium after 2,000 seconds. The numerical results, while similar to those given by Poo, were all somewhat higher than their experimental counterpart. For this series of runs, the diffusion constant was calculated from radius in accordance with Eq. (5.4). The result most like Poo's experimental result was a cell radius of 50 μm, all other cases reaching equilibrium either too quickly (100 μm) or too slowly (10 and 20 μm).

Using the diffusion rate calculated for the 50 μm radius result (5.1×10^{-12} $m^{-1}s^{-1}$), another series of runs was performed varying the exposure level from 100 V m^{-1} to 1,000 V m^{-1}, again to compare with Poo's results. The results shown in Fig. 5.12 are indeed similar to the experimental results obtained by Poo. Once again, however, the asymmetry ratio is numerically somewhat overestimated, although this time, the errors appeared smaller than the previous comparison, the 1,000 V m^{-1} case being the worst case. The lower exposure levels compared reasonably well, comparing time against the asymmetry ratio.

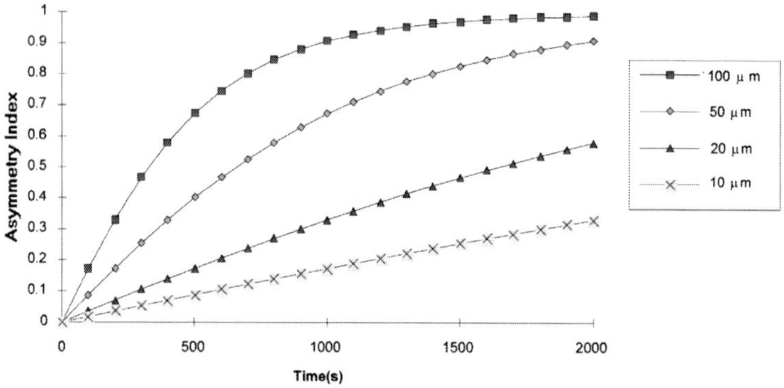

Figure 5.11 Asymmetry index vs. time for a range of cells of varying radii 10–100 μm.

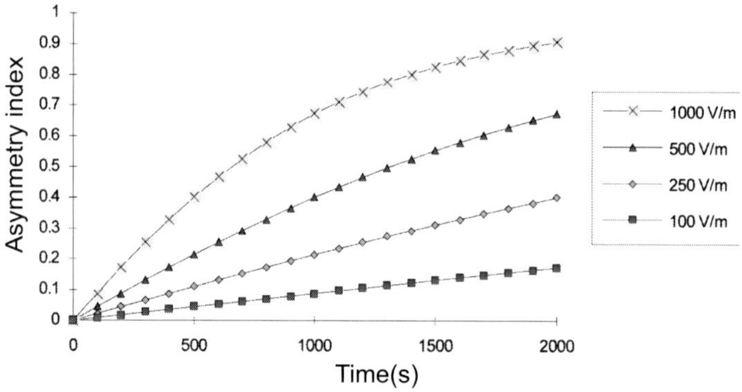

Figure 5.12 Asymmetry index vs. time for a cell of radius 50 μm at differing exposure levels.

Two runs were now made to simulate the experiments of White et al. (1990). Two cell sizes were chosen, 20 μm and 50 μm, while exposure was kept at 20 V m^{-1}. The diffusion constant was as used in the previous run, being set at 4.1×10^{-12} $m^{-1}s^{-1}$. The results show that for the larger cell, equilibrium is reached after about 3 hours, while the smaller cell has not quite reached dynamic equilibrium after 11 hours (Fig. 5.13).

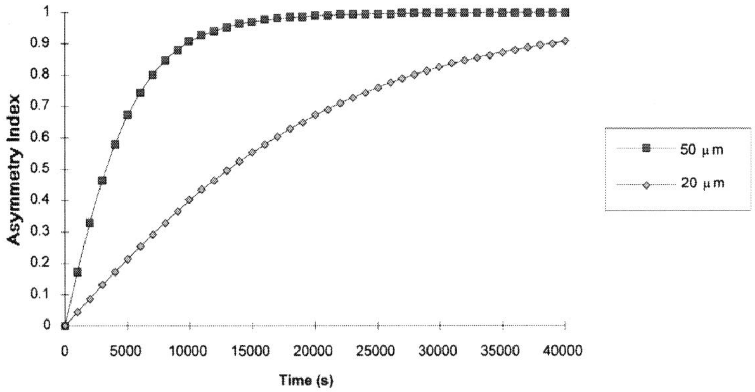

Figure 5.13 Asymmetry index vs. time for cells of radii 20 μm and 50 μm exposed at 20 V m^{-1}.

5.8 Discussion of Results

The fact that higher asymmetry ratios were numerically obtained compared to experimental data may be a result of the underlying assumptions and the actual physics. If we consider that some proteins may be dipolar as mentioned, there may, in fact, be two different classes of diffusing particles. Some diffusive processes, such as those examined by Poo, may be (mainly) due to a net permanent surface charge. Other processes, such as the experiments conducted by White et al., may be (mainly) the result of protein polarisation, where symmetric diffusion might cause a development of foci at both the anode and the cathode, accompanied by a randomly two-sided outgrowth. It should be noted that the dipolar diffusive process would be controlled by the membrane tangential component. As mentioned this membrane component is approximately half the extracellular field just outside the membrane. While the extracellular tangential components drive any permanent net charge on the surface, any tangential intramembrane components would possibly drive a dipole diffusion process within the membrane. Poo's results might be a combination of the two types, so the observation of lower asymmetry ratio is caused by the diffusion of some dipolar proteins along with the diffusion of some permanently charged proteins.

Results obtained from the numerical diffusion code that was developed were found to be reasonably similar to the experimental results obtained by Poo. This indicates that the theory first proposed by Jaffe and Nucitelli (1977) and later expanded by Poo (1981) is a reasonable approximation to experimental observations.

5.9 Cell Growth/Self-Organisation via Constrained Electrophoresis

The growth and development of cells have long been thought to be initiated and controlled by the biogenic electric fields observed around all living cells (Jaffe, 1979; Nuccitelli, 1990; Hush and Overall, 1996). Such biogenic fields are illustrated in Fig. 5.14a,b. In a recent study, protoplasts derived from the filamentous green alga *Mougeotia* were exposed to electric fields of 20 V m^{-1} (White et al., 1990). The protoplasts regenerated displaying a normal sequence of microtubule reorganisation and cell wall deposition. Results indicated the cylindrical cell wall became elongated parallel to the applied field, while the cortical microtubules oriented perpendicular to the applied field. How alignment and growth orientation occur is unclear, but one possibility concerns electrophoresis of membrane components (Poo, 1981).

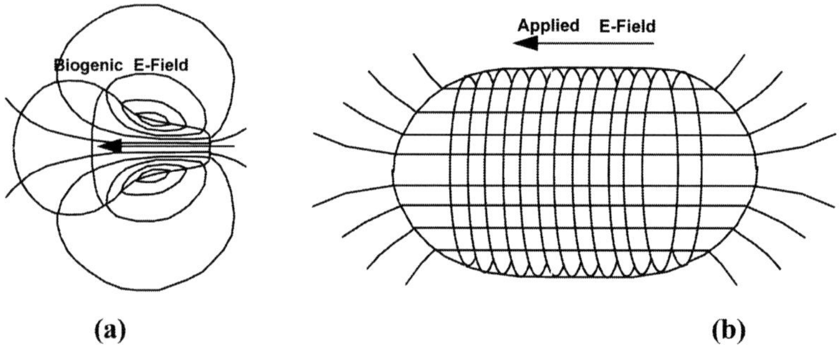

Figure 5.14 Development and growth in cells. (a) Asymmetric growth tip of the seaweed *Fucus* embryo and (b) symmetric regrowth of the *Mougeotia* protoplast. Symmetry is defined in terms of left-right shape. Growth tips are also symmetric top to bottom.

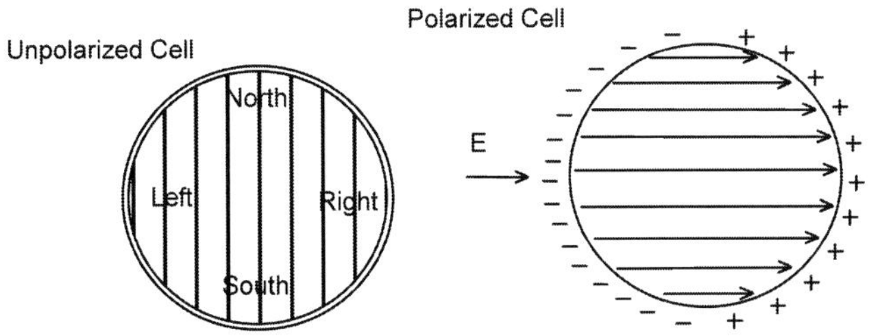

Figure 5.15 (a) Unpolarised cell and membrane and (b) polarised cell showing polarisation arrows.

We now present a picture of protein diffusion that includes both unconstrained *and* constrained membrane proteins, leading to self-organisation. In Fig. 5.15a we have an unpolarised spherical cell or protoplast and its membrane; there is no external E field. Imbedded in the membrane, there are an unknown but large number of proteins that are likewise unpolarised. The cell and the individual proteins are able to be polarised by applying an external E field. This sets up lines of polarisation, as shown in Fig. 5.15b. These lines run parallel to the applied field—see Jackson, pp. 151–154, (1975). The lines of polarisation now affect the distribution of charge in two ways, (1) on the cell surface and (2) within each protein, both on parts of the protein's surface poking out of the membrane (extra- and intracellular) and on those parts that are within the membrane.

On the cell surface, protruding bits of the proteins that were previously unpolarised become polarised by the radial component of the induced E field in the extracellular region, so where there is the need for a negative charge on the cell surface in accordance with the lines of polarisation, the proteins polarise to match this requirement. A similar process occurs within the membrane also due to the strong vertical component of the induced E field. So proteins on the left of the cell shown in Fig. 5.15b polarise, as shown in Fig. 5.16b, while those proteins on the right of the cell in Fig. 5.15b polarise, as shown in Fig. 5.16c. Within the intracellular region, the induced E fields are tiny by comparison. The overall process of induced electric polarisation of membrane proteins is a complicated process by itself, though we will not take further account of this in our model in this report. At this stage we assume the polarisation

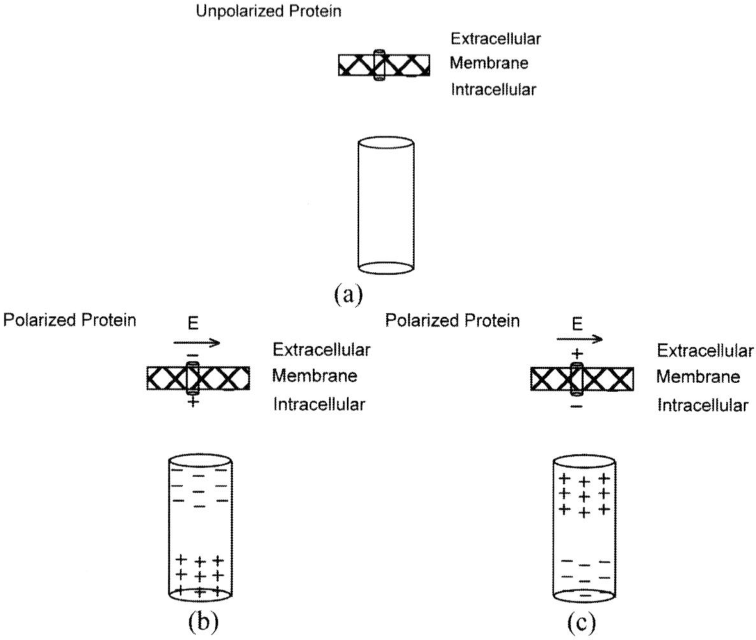

Figure 5.16 (a) Unpolarised membrane protein and (b and c) polarised proteins from the left and right ends, respectively, of the cell in Fig. 5.15b.

process within proteins is fairly rapid in time compared with the protein diffusion within the plasma membrane.

Now, a second, slower phase starts. This is in accord with accepted notions of electrophoresis. Since there is now induced charge separation within proteins, they will start to diffuse. Exactly how this diffusion takes place is not yet clear, but the proteins appear to have a charge running down their length. Perhaps on the extracellular surface there exists a monopole charge on the protruding parts of the proteins. In the presence of an induced extracellular tangential component of the E field these monopoles will diffuse. Inside the membrane, there exists a dipolar charge separation, that is, a dipole, and the induced tangential membrane E field will also cause the proteins to diffuse. Within the intracellular region, the induced fields will probably cause insignificant diffusion compared with the other two components of diffusion.

At the start of the overall process, the distribution and motion of proteins on the cell surface is random. In a time-averaged sense, it is distributed according to Maxwellian statistics while moving under thermal agitation. When the external field is applied, inducing charge separation, as discussed above, the positions of proteins on the cell surface are influenced by components of the induced E field parallel to the membrane, while normal components of the induced membrane E field add to the overall induced dipole moment at these points.

This means that on the polarised left and right ends of the cell, there will be no electrophoretic force, while at the north and south poles there will be maximum electrophoretic force. This is because the induced horizontal E field has no component transverse to the membrane at the polarised ends and is near maximum at the poles (since the surface and membrane E field is a complicated function of geometry, etc.). The proteins congregate near the left and right ends and are pushed away from the poles. At the same time, near the left and right polarised ends, the total induced moment at any point will be large since fields normal to the membrane are maximal at these locations.

If we now look at the theory of arrays of dipoles, we can get the next part of the overall mechanism, and we can see how this is analogous to Frankel's chains of magnetite crystals. In Frankel's case, the crystals are located in chains. 'Normal' dipole array theory, as in Jackson, relates to dipoles that are in each other's 'far field'; the distances separating each dipole from its neighbours, the grid spacing, affect the sum of the dipole moments at each point. In the case of the magnetite crystals, the dipole moments of a chain of *N* crystals contributes to the overall moment by straight addition, whereas we see in constrained theory that the arrays can annihilate the overall moment when the grid spacing is large. We assume that the electrophoretic process has taken place and proteins at the negative end of the cell and those on the positive end are now clustered together and are not evenly spaced left to right, north to south. Observations of tip growth confirm this constrained diffusional process; the formation of tip regions is similar to the formation of a liquid crystal, a regular lattice of proteins that distribute themselves cylindrically around and at the tip. The relative strength of thermal effects when the dipoles form clusters can be estimated as 0.25 kT; in other words random effects are overcome by the self-organisation of the proteins.

In summary the electric field theory of equipotential and field lines within cells and on their membrane surfaces suggests that the formation of ionic currents through cells involves constrained migration of membrane proteins that incorporate suitable transmembrane ion channels. Thus, protein electrophoresis within plasma membranes may help establish asymmetric or symmetric currents and in turn provide an electrical environment for directing asymmetric or symmetric cellular expansion.

In a biographical description of Alexander Gurwitsch's life and work (*Int. J. Dev. Biol.*, **41**: 771–779, 1997) by his grandson Lev Beloussov, the field theory developed by Gurwitsch was outlined. This theory is similar to the diffusive theory given in this chapter and is illustrated in Fig. 5.17a,b for comparison. Gurwitsch's discovery of UV radiations, termed 'mitogentic' radiation, left little time for him to investigate his field theory, but he always kept it in mind.

5.10 Cell Division in Colonies Due to Membrane Protein Dynamics

The next level of the diffusional process involves examination of populations of cells all of which are involved in the electrophoresis of transmembrane proteins, including ionic channels, inherently a multi-particle numerical problem involving many proteins within many cells. Figure 5.18 shows various splines plotted across the frequency spectrum for a generalised high-water-content tissue used in computational dosimetry simulations. At extremely low frequency (ELF), tissues generally exhibit their highest relative dielectric behaviour. This depends on the type of phenomenon we have been examining involving the formation of ionic currents across and transmembrane proteins within the plasma membranes of many cells acting in cooperation. This is a form of cell–cell signalling, whereby all cells act together to form an enhancement of the initial signal. The initial signal may come from a variety of causes, including chemical cascades, for instance, a wound, or

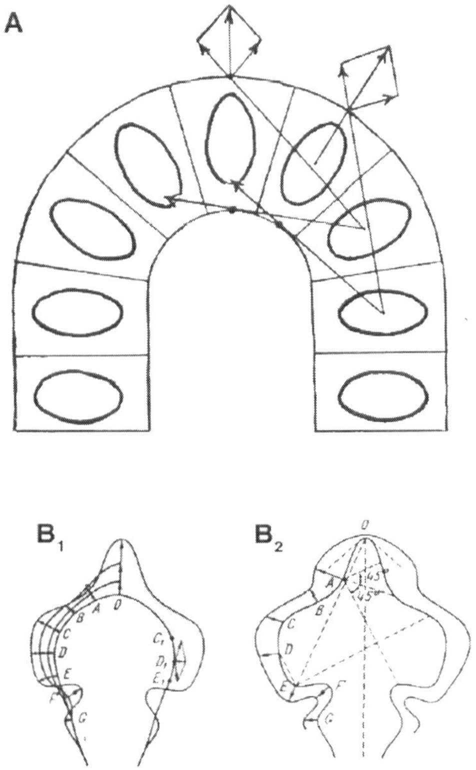

Figure 5.17 Illustrations of the 'curvature-increasing rule' based on the 1944 version of Gurwitsch's field theory. (A) Field vectors at the tip of archenteron invagination of the sea urchin (Gurwitsch, 1944, Fig. 23). (B1 and B2) Application of the curvature-increasing rule to the morphogenesis of the hydroid polyp *Obelia loveni*. In (B1) the shift of points 0 and A–G on the initial contour are assumed to occur along a path perpendicular and proportional to the original contour of the curve. In (B2), the interactions between the opposite walls of the primordium are introduced (dotted lines) in a manner suggested by Gurwitsch (1944). As a result rather realistic changes of shape are modelled (Belousov, 1968). Later, another model for the same changes of shape was introduced (Beloussov and Laklrev, 1991). (*Source*: Beloussov L.V., Life of Alexander G. Gurwitsch and his relevant contribution to the theory of morphogenetic fields, *Int. J. Dev. Biol.*, **41**:771–779, 1997; additional commentary by J.M. Opitz (Department of embryology, Faculty of Biology, Moscow State University, Moscow, Russia), S.F. Gilbert (Pediatrics, Human Genetics and Obstetrics and Gynecology, University of Utah, Salt Lake City, UT), and Department of Biology, Martin Biological Laboratories, Swarthmore College, Swarthmore, PA.)

a developmental need within a cell cycle. In this second case a cell nucleus is assumed to have dissolved, releasing its contents, including the constituent elements of the spindle poles that are charged and form microtubule attachments to the inside of the membrane of some central cell. This process within a central cell forms the biogenic electric field, causing the electrophoresis of membrane proteins in a population of neighbouring cells. In these cases very large enhancements can occur, for example, the liver ($\varepsilon_{\mathrm{r}}^{\mathrm{liver}} \approx 10^7$).

The speed of the overall division process can vary; it is known fo,r instance, that in general malignant cancer cells proliferate very rapidly relative to normal healthy cells. This fact can be used to selectively destroy melanomas in the treatment using surface dipoles devised by Kirson et al., as discussed in Chapter 1.

A related effect to this picture of cell polarisation leading to large enhancement of the initial signal occurs when a magnetic effect is also present. The observation of dispersion levels across the spectrum is thought to be due to various biological interactions, including magnetorotation of cells. Dispersion levels involve components of various sizes increasingly getting smaller as frequency grows. Smaller and smaller time constants are involved. At the ELF, α-dispersion, it is not only the size and the time constant that determine the dispersion magnitude but also the enhancement phenomenon of cell–cell cooperation across populations of cells that are undergoing membrane electrophoresis that cause ionic current flows within the cell. The constrained membrane diffusion theory discussed above is connected to the well-known voltage-gated transmembrane diffusion theory of Hodgkin and Huxley with its on-gate potentials typically around 65 mV where cells are activated from a resting state.

In a recent study static H fields of varying size 10, 14, 18 and 22 mT were shown to have an effect upon the dielectric response of rat liver. The rats were whole-body-exposed to these intensities, one hour daily for one week. The results showed noticeable variations in normal values of, σ of liver tissue after exposure to all H field intensities. This demonstrates the link between static H fields and electrophoresis, as with other experiments where magnetorotation of cells has been observed.

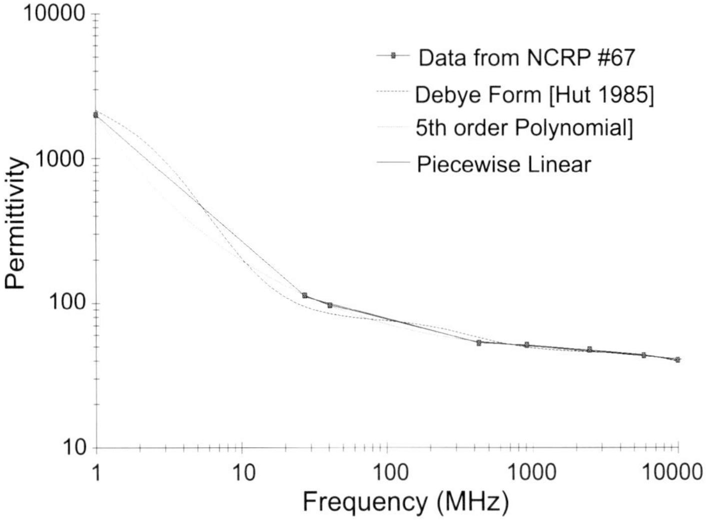

$$y_1 = 3.300987 - 1.47071x + 0.609729x^2 - 0.17636x^3 + 0.035895x^4 - 0.00339x^5$$

$$y_2 = \frac{2.34 * 10^7}{\{1 + (f/78)^2\}} + \frac{6.21 * 10^4}{\{1 + (f/76 * 10^3)^2\}} + \frac{1.97 * 10^3}{\{1 + (f/2.6 * 10^6)^2\}} + \frac{30.8}{\{1 + (f/3.4 * 10^8)^2\}} + \frac{41.3}{\{1 + (f/2.3 * 10^{10})^2\}} + 4.3$$

$$y_3 = \begin{array}{ll} 8.5251 - 0.87068x & \text{where } 1\text{ MHz} < f < 27\text{ MHz} \\ 4.0342 - 0.26763x & \text{where } 27\text{ MHz} < f < 433\text{ MHz} \\ 2.5023 - 0.08920x & \text{where } 433\text{ MHz} < f < 10000\text{ MHz} \end{array}$$

where $y_1 = \log_{10}(\varepsilon_r)$

$y_2 = \varepsilon_r$

$y_3 = \log_{10}(\varepsilon_r)$

and $x = \log_{10}(f)$

Figure 5.18 Various splines for interpolating permittivity of high-water-content tissue.

Turning to the situation as metaphase approaches, the chromosomes are being pulled by the formation of ionic currents through the north and south poles of a cell, as in Fig. 5.18a–c. As this process proceeds the forces pulling the chromosome get stronger as diffusion of proteins within the membranes of the surrounding colony of cells proceeds. As the electrostatic forces get stronger and stronger (Gagliardi, 2007) the helices of the chromosomes (in

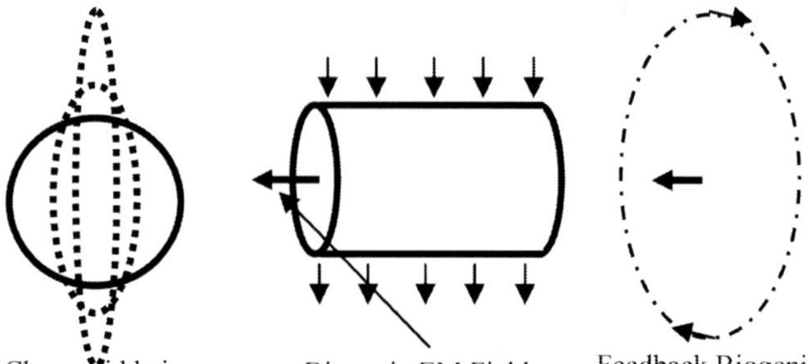

Figure 5.19 Chromatid being stretched vertically. (a) Cross-section shown at three stages by the joint actions of the developing biogenic feedback E and H fields shown in Fig. 5.20a–c. The E field induces flow of biophotons along the axis of chromosomes (one chromatid illustrated in (b) being impinged by the E field transmitted via microtubules from spindles). (c) The biogenic H field depends on the flow of biophotons due to the stretching and results from the change in shape of the cross-section of the chromatid at or near metaphase.

particular the chromatids) may change their overall shape from circular to elliptic, as shown in Fig. 5.19a.

The electrostatic forces induce a change within the atomic structure. An induced EM radiation forms within the loops of the chromosomes. Termed the Stark effect, when an atom is exposed to an external electrostatic field this causes a shifting and splitting of spectral lines of EM radiation emitted by the atom due to its exposure to the static E field. In terms of its dynamics the orbit of the outer shell electron becomes an ellipse. This shift in atomic energy level causes the emission of a photon with a specific energy by the atom. This appears to be the cause of the biophotons observed by Popp et al., who have used photomultipliers to count small numbers of biophotons emitted by strands of DNA when exposed to various forms of radiation.

Due to the helical loops within the molecular structure of the stiffening chromosomes, this radiation forms a flow of biophotons down the axis of the chromosome. This process may not involve the entire length of the chromosome, but as diffusion of the colony proceeds, and the biophotonic flow within the loops of the chromosomes grows larger, the biophotonic flow grows larger forming a biogenic EM field.

Thus at metaphase the chromosomes are assumed to be pulled on each side, north and south, by the formation of ionic currents through the north and south poles of a cell, as illustrated in Fig. 5.20a. We assume a colony of cells has enhanced the ionic current flow through the cell to bring about the stretched state of the chromosomes. These stretched chromosomes release biophotons down the axis of their helical structure. This forms an axial current that in turn creates an H field pointing out from the helical axis. In turn this magnetic field may cause magnetorotation within cells in the east–west parts of the colony as distinct from the north–south parts of the colony experiencing electrophoresis. This magnetorotation effect may enhance the magnetic field, reducing the ambient energy density at the site of the chromosomes.

Figure 5.20b illustrates the type of cellular magnetorotation that would help explain some aspects of division. What is not clear is the role any biomagnetism plays in the interacting systems, chromatids, histones and membrane proteins. The histones within the DNA may play a role in the suggested magnetorotation process. Observations of intense agitation and stirring somewhat like clouds before a storm are indicative of a process involving magnetorotation. What may be occurring is a rain-like process in which the spine of hydration surrounding the chromosomes breaks apart causing the onset of anaphase. Gibbs free energy is reduced by the magnetorotation of the neighbouring cells.

In 2001 Van Wijk, reviewing Gurwitsch's mitogenetic fields, commented:

> The research on bio-informational aspects of bio-photons in the IR to UV range can be traced back to Alexander G. Gurwitsch more than seventy years ago. He emphasized that fundamental biological functions such as cell division are triggered by a very weak ultraviolet photo-current originating from the cells themselves. This postulate of bio-photonic information appears to many scientists to be pure speculation, and it provokes sometimes contempt rather than carefully considered objections. This article reviews the activities of research groups on three different main

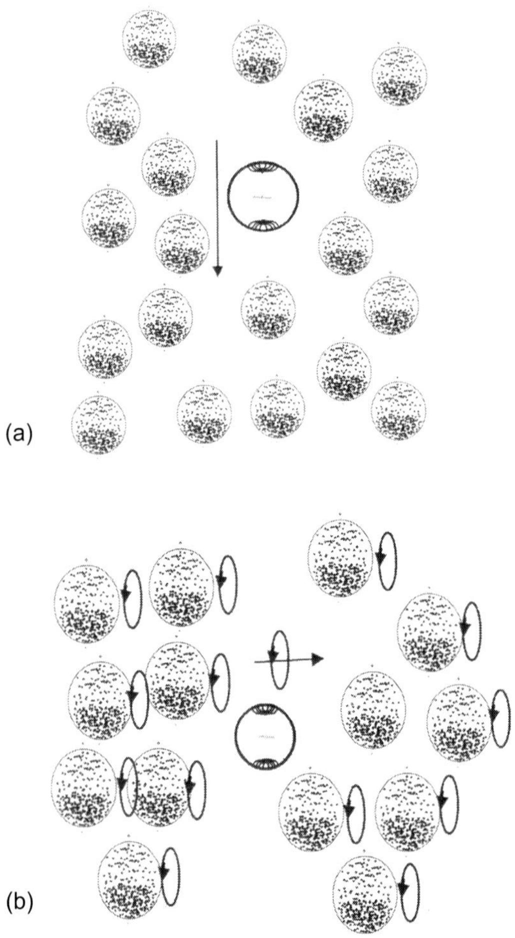

Figure 5.20 (a) Colony of cells aligning and polarising to create a biogenic E field through the north–south axis of a central replicating cell whose spindle poles have moved apart to the north–south poles after the cell nucleus melts to initialise the E field in the cytoplasm of the central cell. (b) Colony of cells rotating to create a biogenic H field across the east–west plane of a central replicating cell that is emitting a flow of biophotons down the axis of the chromosome of the central cell, thus reducing the energy within this central cell, enabling metaphase.

questions concerning this bio-photonic information. The first question deals with the developments in the evidence for photons originating from cells. Despite serious experimental difficulties it is now clear to every scientist working in this field that photon emission could be detected from nearly all living cells.

The second question considers the origin of photon emission. Very weak photon emission has been looked upon so far mainly from the possible reactions and biochemical pathway that could be responsible for this phenomenon. In general, those studies were carried out without considering Gurwitsch's idea of bio-information of photon emission. An alternative search for the origin of photon emission has been carried out incorporating the informational aspect of photon emission. This type of explanation proposes the existence of a coherent electromagnetic field within cell populations and has led to the introduction of the term bio-photons. Bio-photons are characterized by their quantum character and are supposed to escape from a coherent field. This alternative explanation is supported by several arguments.

The third question is the most decisive one from an empirical point of view. It is directed to the existence of bio-photon emission in relation with cellular interactions and biological function. In general the idea that beside, or even below, the biochemical level of control very weak electromagnetic interactions play a regulatory role in the living state has received relatively little attention. The present research has not yet reached the state required for the ultimate verification or falsification of the hypothesis on bio-photonic information in cell division and other cell physiological processes, as originally investigated and suggested by Gurwitsch.

The mystery of the sporadic arising of cell divisions was the starting point for Gurwitsch to carry out his famous 'mitogenetic radiation' experiments in 1923. The idea that radiation generates cell division was based on his early studies (Gurwitsch, 1911) in which it was demonstrated that (1) there is a reverse linear relationship between the surface areas of the meristemic cells and their division frequencies; (2) along the whole onion root meristem, cell surface areas increase according to the exponential law. The purely statistical character of spatial distribution of mitosis demonstrated in several objects (and particularly in onion roots) supported the concept that mitosis should be based upon a dual principle. That is, one of the factors which make a cell capable of division is assumed to be endogenous (a 'possibility factor'),

> while a second one ('realization factor') is exogenous although it may arise in the same organism.
>
> These early experimental observations were interpreted in the following way: there exists a surface 'principle K', which remains constant during cell growth, and also a 'principle A' which increases in a metabolic manner (Gurwitsch, 1922). The main problem was to elucidate the nature of the exogenous principle. Initially it seemed natural to look upon it as a chemical substance.
>
> However, cell division frequencies should then be proportional to the relation of K to a constantly increasing A, which contradicts the established fact of a reverse linear relationship between cell division frequency and cell surface area. This contradiction led to the following suggestion: K and A sites are arranged on a cell surface as a permanently changing spatial mosaic. It is the mosaic-like configuration which plays a decisive role, the perception of an exogenous impulse by a cell surface may be considered as a resonance event. This led to the hypothesis that the exogenous division stimulating principle is not a chemical substance, but instead an oscillation process which may be a radiation.
>
> The experimental verification of the hypothesis that the exogenous factor is a form of radiation has evolved from the suggestion that at least some radiation should emanate from root cells into the surrounding space and that it would be most probable that detectable radiation would arise from the cone-shaped tip of an onion root. An adequate detector should consist of a second root with cells ready to divide and really dividing with a certain average frequency, but at the same time capable of increasing the frequency. The necessity for comparison with control cells was also obvious. In that respect onion roots are suitable objects due to their radial symmetrical arrangement. Therefore, the revealing of a difference in mitotic numbers between the 'irradiated' and the 'shadowed' sides of the meristem. . .

We can examine the situation in qualitative fashion. The mean surface area A of a colony of approximately spherical cells is $A = 4\pi r^2$, where r is an averaged cell radius. If charged proteins within the membranes of nearby cells in the east–west direction (perhaps many such cells) are performing magnetorotation with an average speed of $v = \omega r$, then the mean surface area A of these cells relates to the magnetic or rotational energy emitted by a stretched chromosome of a central fertilised cell. This magnetic energy relates to the flow of biophotons within the chromosomes and to the energy of the E field at the division $E = h\nu_R$, an SFT cellular balance. As we saw in Chapter 4 a cell exposed to an EM field responds via rotations within its membrane. This is the link between the biogenic E and H fields illustrated in Fig. 5.17. The chromosome eventually reaches the limit of its states and can no longer sustain further stretching. While chromosomes have a similar structure across the body the biological milieu varies, yielding differing replication frequencies ν_R. This mechanism is based, in part, on the EM field generated from the distortion within the chromosome that occurs due to the cooperative alignment of cells, leading to metaphase. The subsequent cellular rotation[a] effect leads to anaphase. Thus there is a combined electric and magnetic effect typical of the SFT formulation.

One last point to be made concerning acoustic (A) fields is as follows. A similar effect to metaphase-anaphase occurs when large electrostatic fields build up between rain clouds and the earth to act on the clouds. A similar stretching may occur to create the H fields and turbulent winds that then appear (see Fig. 5.21). Eventually thunder is heard, and rain follows, falling to the earth shortly afterwards. Where does the acoustic energy of the thunder come from? Where there is a significant temperature changes in a region it may be that the nuclear state changes, inducing a concomitant change in the nuclear binding energy. If the nuclear energy state and the binding energy change within a region then the E, H and A fields will all change simultaneously. Hence wind, thunder and rain become a simultaneous occurrence as the nuclear state changes. We would expect that if the above nuclear effect occurs, the electric, magnetic and acoustic energies might be related in a stoichiometric fashion or perhaps be even equal to each other. We point out that

[a]This rotation effect is due to an EM field. While the effect is similar to magnetorotation, the effect is also similar to a gravitational field between dipoles, although in this case, as explained in Chapter 4, the cell's membrane, a structure consisting of myriads of dipoles, is being exposed to mono-polar EM sources due to any cells at or near metaphase. To be pedantic we should refer to this effect as an EM effect, but like the spin of the earth on its axis, the effect is indeed magnetic-like, a part of an overall coupled EM effect.

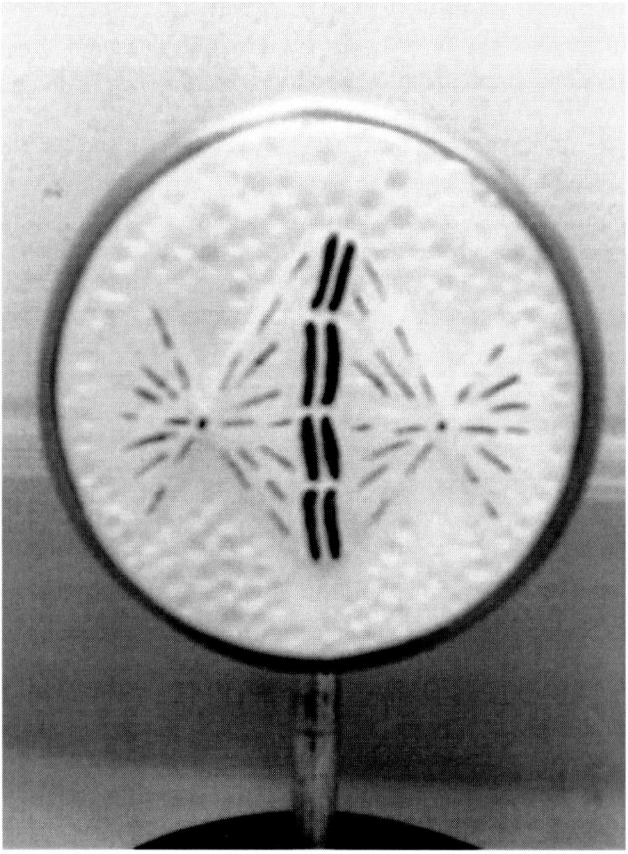

Figure 5.21 Model of metaphase (*source*: http://faculty.irsc.edu/faculty/jschwartz/Cell%20Cycle.htm).

there may be similar conditions within the interaction of the EM forces and the acoustic forces (i.e., not within the nuclear regions). In the next chapter an acoustic (or vibrational) device is premised upon the same energy for a bond length in DNA to that of a corresponding EM wave. Hence acoustic energy may be related to the changes in the nuclear states of DNA during mitosis and metaphase in equal measure to the EM field. According to Bauer cells do indeed emit an acoustic frequency during mitosis. As noted by Fleming, an acoustic 'spectroscope' (and indeed a magnetic 'spectroscope') may be a viable instrument to study such near-field effects. The use of an acoustic device, a modern, perhaps computerised form of the stethoscope, may form a critical aid for the modern physicist, general practitioner and oncologist.

Chapter 6

Electromagnetic and Acoustic-Based Therapies

Medical science has begun to embrace the general modality of using non-ionising energies of static E and H fields and electromagnetic (EM), ultrasonic and audible acoustic (A) fields as various means of delivering therapies. There is a critical need to improve current techniques and to engineer methods for many diseases where there is a paucity of current effective surgical or pharmaceutical treatment, such as critical tissue cancers, motor neuron disease, multiple sclerosis and Alzheimer disease. Novel areas of medical application may include neuroscience where problems such as addiction and psychosis are common aspects of modern life. Medical science is now engaged in the battle for health, incorporating groups across the globe from the East and the West. Therapeutic methods incorporating static and time-varying fields use applicators either directly to the skin or interstitially, or as radiations, to generate waveforms of differing forms of energy that can be applied to organs, tissues and anatomical areas of the body. In this chapter we examine two methods with differing characteristics and mechanisms. The first using EM fields induces biochemical cascades where the EM field acts as first messenger, which piggy-backs on to the body's

---

This chapter was co-authored by Elizabeth Bauer.

Inside the Photon: A Journey to Health
Tony Fleming

ISBN 978-981-4241-40-3 (Hardcover), 978-981-4241-88-5 (eBook)
www.panstanford.com

natural methods of growth and repair. Non-thermal levels of pulsed EM fields (PEMFs) can be applied via interstitial electrodes to induce significantly enhanced concentration levels of Ca^{2+} acting as the second messenger on to the surface of cells in the chemical pathway associated with regeneration. Non-sinusoidal, for example, pulsed, waveforms create field components larger than the sinusoidal case. Therapeutic, medical and post-surgical benefits are observed. In the second case a spatially variable H field is used in conjunction with a frequency-variable A field. This method uses the penetrative powers of vibrations relative to EM fields within tissues of the body. The fields are applied to the skin via an applicator incorporating a permanent magnet. The applicator and magnetic head vibrate at a frequency under therapist control. A range of results are presented as a series of before and after thermographs and ultrasonographs.

6.1 Introduction

We have seen that photons are involved in both the big picture of life and evolution within our galaxy and across the cosmos and in the small picture of the binding energy within atoms and molecules, including acting as initiators of sudden changes of field structure, chemical cascades during mitosis or growth of tissues due to biogenic field exposures, biophotons (see Fig. 6.1). Using classical electromagnetics (CEM) the role of the biophoton in cellular replication was unable to be understood. CEM does not include the biophoton and how it acts within atoms and molecules in the body's mechanisms in any detail. However, frequency-based methods of therapy are such an intrinsically simple resonance mechanism for treating cellular diseases, including cancers, they do not need the recently discovered mathematics and physical concepts of self-field theory (SFT) to understand how they work. But the mathematics certainly helps in validating the results to the academic community. This validation refers to the way molecular, chromosomal and cellular changes come about in response to the photonic and phononic exposures.

Hence frequency-based methods of therapy have been known and attempted since at least the early 20th century. Gurwitsch

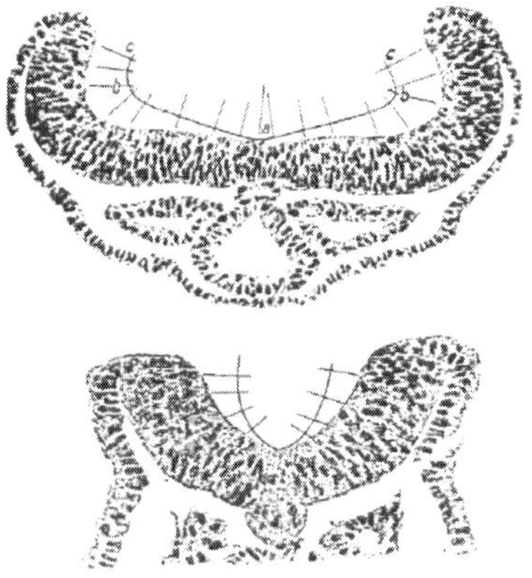

Figure 6.1 'Prognostic' orientation of longitudinal nuclear axes in a selachian neural plate. (Top) The prediction based on theory as to where the surface of the neural plate will move morphogenetically is represented by the line cbac, the contour of the line being based on the lines drawn perpendicular to the surface of the neural plate through the long axis of representative nuclei in corresponding mirror-image sides of the neural plate. Exactly this line will be reached in about 1 h of further development. (Bottom) Predicted surface of the neural plate at a somewhat later stage of selachian neural tube formation. (Credit: Beloussov L.V., Life of Alexander G. Gurwitsch and his relevant contribution to the theory of morphogenetic fields, *Int. J. Dev. Biol.*, **41**:771–779, 1997.)

observed a link between division frequency and the surface areas of cells. This provides a connection to the possible SFT mechanism at metaphase involving biophotonic emissions from the chromatids being pulled apart as part of the cell cycle. Royal Raymond Rife initially invented what he called a dark-field microscope, and the work in understanding how light resonates at microscopic levels led him to understand how resonance could be used at a biological level as a therapeutic aid (see Fig. 6.2). Unfortunately the scientific reality in the early part of the 20th century was that such mechanisms were a long way from being understood. There are interesting synergies between the work and travels of Gurwitsch from Poltava in the Ukraine and Rife from an ethnically German region of Nebraska in the U.S.; both travelled and worked abroad extensively, visiting Germany prior to the end of World War I (WWI). The application of their work was directed towards the medical treatment of cancer;

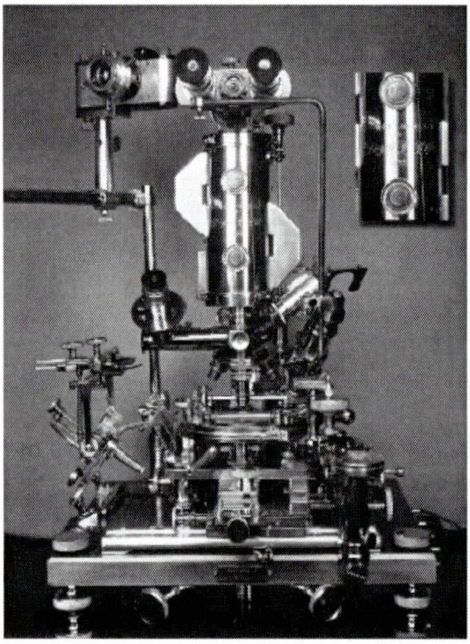

Figure 6.2 Rife universal microscope #3.

both had the joy of a golden era of acceptance from the 1920s into the 1930s, only to suffer a subsequent period of rejection just before and in the decades after World War II (WWII). The work of both is now being reassessed in the light of recently designed therapeutic methods closely related to their pioneering efforts nearly a century ago.

At the start of the 21st century resonance mechanisms are intrinsically very simple to understand. These resonance methods may be classed as either destructive or constructive, similar to the nomenclature of resonance and anti-resonance, where waves can cause destructive or constructive interference, depending on the particular resonance type. Where the phase length of the oscillator is a multiple of π the wave is constructive, and where it is a multiple of $\pi/2$ the wave is destructive.

As one possible method to treat cancer, deoxyribonucleic acid (DNA) can be attacked at metaphase when its length, or alternatively its width, is extended somewhat and has a rigid liquid crystal structure. This is a destructive use of resonance, where the frequency is chosen to match the length or width of the chromosome. Often, cancers proliferate at higher rates than normal cells. We might choose a part of the body where the cancer is somewhat isolated from critical organs and tissues and where a lack of replication of normal tissue is not going to be problematic. For instance, a cancer on the arm may be attacked in such a destructive fashion. When the chromosome reaches metaphase it will react to sufficiently high field levels of electromagnetic fields (EMFs) so that the structure becomes unstable. The advantage of acoustic or vibrational frequencies over electromagnetic (EM) frequencies is that EM waves are attenuated more by the tissues of the body and it becomes a problem to generate sufficient power within the tissues of the body. Microbial pathogens inside the body can also be attacked via their characteristic length. We can also use a similar method based on constructive wavelengths to treat healthy cells. Frequencies associated with stages of the cell cycles of the various organs, tissues and cells can be promoted via their known biophotonic frequencies. In the battle to overcome disease and ailments, any advantage that can be applied is a potential therapy. The use of methods to maintain and prolong health can be used as an aid to overcome unhealthy tissues and organs due to many diseases.

6.2 Pulsed EMF Therapy

Over the past decades some of the proposed mechanisms have evolved to become real methods in the hands of the clinician. This includes the windows mechanism already described. Markov, for one, has written extensively of the various bioelectromagnetic (BEM) devices available at present:

> Magnetic and electromagnetic fields are now recognized by the XXI century medicine as real physical entities that promise healing of various health problems, even when conventional medicine has failed. Today magnetotherapy provides a noninvasive, safe and easy method to directly treat the site of injury, the source of pain and inflammation, and other types of diseases and pathologies. Millions of people worldwide have received help in treatment of musculoskeletal system, as well as pain relief. Pulsed electromagnetic fields are one important modality in magnetotherapy and recent technological innovations, such as Curatron PEMF devices, offer excellent, state of the art computer controlled therapy system. . . .

Some examples for target populations

The largest populations of patients that have received, or could benefit from magnetic field therapy are victims of musculoskeletal disorders, wounds and pain. Following is a summary of information for the number of people in the USA who needs help in abovementioned areas. Five million bone fractures occur annually in the United States alone. About 5% of these became delayed or nonunion fractures (Ryaby, 1998).

According to National Osteoporosis Foundation about 10 million Americans have osteoporosis and 34 million(s) of US citizens have low bone density, which put them at risk for further musculoskeletal disorders. Chronic wounds and their treatment are an enormous burden on the healthcare system, both in terms of their cost ($5 billion to $9 billion annually) and the intensity of care required. There is even more cost to society from human suffering and reduced productivity. More than 2 million people suffer from pressure ulcers and as many as 600,000 to 2.5 million more have chronic leg and foot wounds (Wysocki, 1996). Diabetic foot ulcers are probably the most common chronic wounds in western industrialized countries. Of the millions who have diabetes mellitus, 15 per cent will suffer foot ulceration which often leads to amputation 100,000 per annum in the US alone). (Pilla, 2006) The National Institutes of Health estimate that more than 48 million Americans suffer chronic pain that results in a 65 billion loss of productivity and over $ 100 billion spent on pain care (Markov, 2004c). Better part of this money is spent for pain - relief medications. Recent advances in magnetotherapy suggest that carefully selected magnetic fields might be helpful in treatment of diseases as Parkinson's, Alzheimer, as well as Reflex Sympathetic Disorders which have relatively small number of potential users.

(*Source*: Markov M., Pulsed electromagnetic field therapy, history, state of the art and future, submitted to Electromagn. *Biol. Med.* for publication Nov. 2006.)

Again quoting from the *Journal of Bone and Joint Surgery (Br)*, Hannouche et al., in a review article of pulsed electromagnetic fields (PEMFs), wrote in 2001:

The development of alternative techniques to treat nonunion or bone defects offers promising perspectives for a large number of patients. Physicians treating fractures will have an increasing variety of options available, some of which will probably soon supersede harvest of bone autograft. Physical methods are non-invasive and have been shown to offer beneficial effects in the healing of fresh fractures (mechanical loading, ultrasound stimulation), and in the treatment of hypertrophic nonunion and congenital pseudarthroses (electromagnetic fields). These methods may be useful clinically, especially since some have been shown to reduce substantially the costs of treatment. 75 More recently, the development of techniques of tissue engineering which aim to construct bone tissues in the laboratory, have shown great promise for the replacement of massive bone defects. These techniques are invasive and expensive, however, and further problems must be overcome before they can be adapted for widespread clinical use. The long-term effects of implanted growth factors must be carefully evaluated, and suitable delivery systems developed to provide the slow release of active molecules which can target specific areas of the construct. The enhancement of fracture healing is not only a major issue of health-care, with potential benefit for millions of patients every year, but is also a potential market for the industry. Further independent research must be conducted to make quantified measurements in vivo, in animal models and in controlled clinical trials undertaken to ascertain the role of these techniques in improving the healing of fractures with identification of the most effective material.

(*Source*: D. Hannouche, H. Petite, and L. Sedel, Current Trends in the Enhancement of Fracture Healing, The Hopital Lariboisiere and the Laboratoire de Recherches Orthopediques, Paris, France, http://web.jbjs.org.uk/cgi/reprint/83-B/2/157.pdf.)

6.3 Acoustic/Magnetic Treatment

Perhaps the most promising treatment involves the use of acoustic (A) or vibrational fields. Bauer currently treats many different illnesses and diseases, building up a repertoire of treatment protocols, including frequencies in each different therapeutic case.

While PEMF is much more diffuse amongst the scientific community, the acoustic therapy community is a growing and thriving group of different clinicians whose use of acoustics ranges from psychologists to musicians to those like Bauer, who use an applicator whose frequencies can be modified on a case-by-case basis. The following case study presented at the Bioelectromagnetics Society meeting in Dublin in 2005 gives one such example, where a thoroughbred horse was treated for a bone fracture with excellent results. Note that the magnetic (H) field aspect of the CYMA 1000 was not discussed. What may be happening to affect therapy is that vibrational energy is able to penetrate the tissues more easily than a corresponding EM wave.

Abstract: The case study aimed to assess the effectiveness of an acoustic device on core tendon lesions of the thoroughbred race horse. Electromagnetic self-field theory (EMSFT) shows how the photon and the phonon have a similar sub-structure but an orthogonal orientation to each other. Hence sound, as well as EM exposure can interact to cause cellular effects within the cell-cycle such as replication or apoptosis. The therapeutic method involved the application of an acoustic device (Cyma 1000), which delivered specific frequencies, within the audible sound range, to acupuncture points and meridians as well as the areas of injury. These audible frequencies ranged from 100 to 1600 Hz. In our case studies, we have found a rapid rate of tendon healing; as evidenced by tendon healing with ultrasonographic imaging that clearly showed the return to normal and homogeneous tendon cell integrity as proven by the uniform and normal echogenicity. A complete reduction of the core lesions to a well defined tendon cell regeneration area with no evidence of prior tendon lesions was found. The mean duration of healing time, per follow-up ultrasonography, as well as the clinical signs of healing and rapid return to function was 40 days; from the first diagnostic ultrasound scan to the last follow-up ultrasound scan. The case studies demonstrated a significant improvement in pain and swelling, as well as objectively measurable functional improvements. Objective markers such as clinical improvement in ROM, function and return to work with no signs of lameness at the jog or the gallop. With the encouraging results of these case studies, further investigation of the efficacy of this acoustic device is warranted. The high quality of the homogeneous, healed, tendon tissue, per ultrasonographic evidence, translated into a low to null continued morbidity or threat of reinjury. The cost savings in veterinary diagnostics, care, treatment and maintenance over time, as well as the preservation in value of this athletic horse will further insure this horse owner a gainful return on their substantial financial investment.

Theoretical Considerations

Theory: EM self-field theory (EMSFT) [Fleming 2005] is an analytic solution to Maxwell's equations. Self-field theory (SFT) provides solutions applying to various physical phenomena including a generalization of Maxwell's equations suitable for the nucleus based on a photonic chemistry [Fleming and Colorio (nee Bauer) 2004a, 2004b]. Quantum field theory (QFT) and SFT taken together provide an enlarged vision of physics; both QFT and SFT are valid, producing similar but essentially different perspectives of the one physics. An analogy to QFT and SFT lies in the numerical techniques known as finite element method (FEM) and finite difference method (FDM). FEM uses Lagrangians involving integrations and FDM uses partial differential equations directly to solve systems of equations; FDM is numerically and analytically simpler than FEM. Similarly, SFT applies solutions to the Maxwellian divergence equations into the remaining curl equations directly, rather than using Hamiltonians etc. SFT thus provides a basis for a unification across physics and biophysics.

Associated with changing energy levels across the cell cycle, there are structural changes originating within the fields of proteins such as the DNA that correspond to transition shifts in energy within each cell. Generally DNA can be controlled via delivery of energy of appropriate discrete frequencies involved in the cell cycle. One way to achieve this is via externally applied photons of precise frequency, i.e. EM radiation applied externally to any part of the body, in this case, to the tendons of thoroughbred horses. Another associated technique is to use acoustic energy instead of EM. In EMSFT, the photon and the phonon are theoretically shown to be two particles having the same structure, but a different propagation vector. The phonon is orthogonal to the photon of equivalent energy. The phonon has the important advantage of being able to penetrate far deeper into biological tissues than EM radiation. Other acoustic interactions also occur inside atomic nuclei where gluons, photons, and phonons can interact. The various methods of energy delivery to either promote cell replication or cell apoptosis are generic to a wide class of medical therapies.

Intervention

The application of an acoustic device (Cyma 1000), which delivered specific frequencies, within the audible sound range, to acupuncture points and meridians as well as the areas of injury. These audible acoustic frequencies ranged from 100 to 1600 Hz.

Objective: To assess the effectiveness of an acoustic device on core tendon lesions of the thoroughbred race horse.

Outcome: In our case studies, we have found a rapid rate of tendon healing; evidenced by ultrasonographic imaging that clearly showed the return to normal and homogeneous tendon cell integrity as proven by the uniform and normal echogenicity. A complete reduction of the core lesions to a well defined tendon cell regeneration area with no evidence of a prior tendon lesion was found.

The mean duration of healing time, per follow-up ultrasonography, as well as rapid return to function was 40 days; from the first diagnostic ultrasound scan to the last follow-up ultrasound scan. All subjects, thus far, have had very similar results, therefore, only one complete case study will be presented here.

Primary Diagnostic Ultrasonographic Images

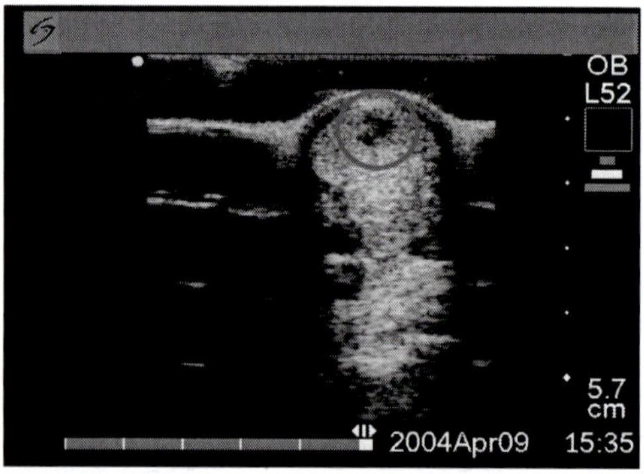

Image 1 Short axis ultrasonography of the SDFT 4-09-04.

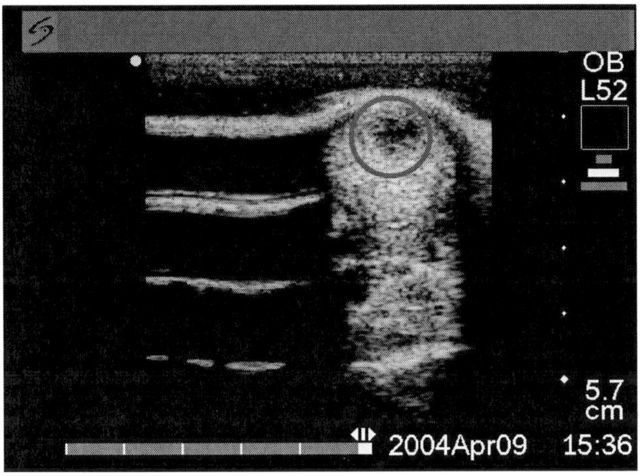

Image 2 Short axis ultrasonography of the SDFT 4-09-04.

Ultrasound scan images during this primary clinical evaluation showed a very concise 25 to 30% core lesion of the superficial digital flexor tendon (SDFT).

First Follow-up Ultrasonographic Image

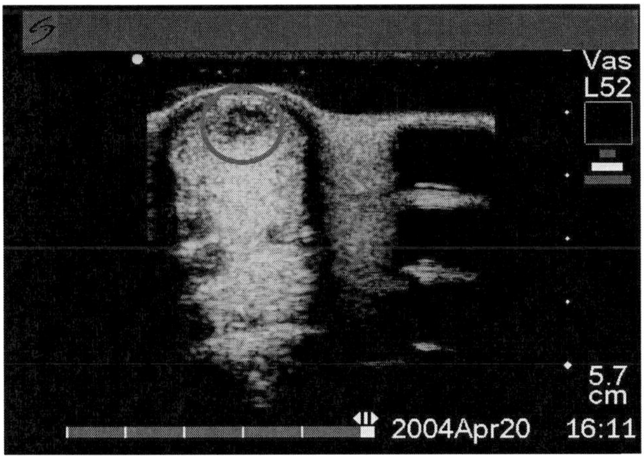

Image 3 Short axis ultrasonography of the SDFT 4-20-04.

Clinical Evaluation

The first follow-up Ultrasound scan showed a 50% improvement in the SDFT core lesion size, with tendon cell density, fibrin deposition and collagen fibril formation as a clear indication of the hypoechogenic healing of this SDFT tendon lesion.

Clinically, there was a decrease in lameness from a Grade 3.5 to Grade 2 of 5. There was decreased sensitivity to flexion and palpation of the flexor tendon in the left front limb. There was also a marked decrease in swelling and no more heat felt on palpation to the tendon area or the limb.

Second Follow-up Ultrasonographic Images

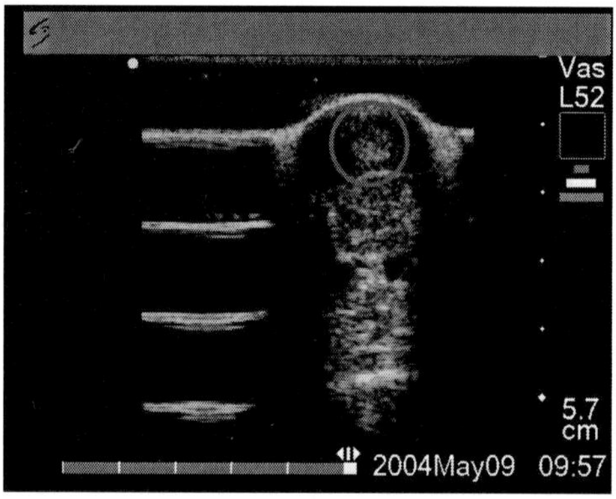

Image 4 Short axis ultrasonography of the SDFT 5-09-04.

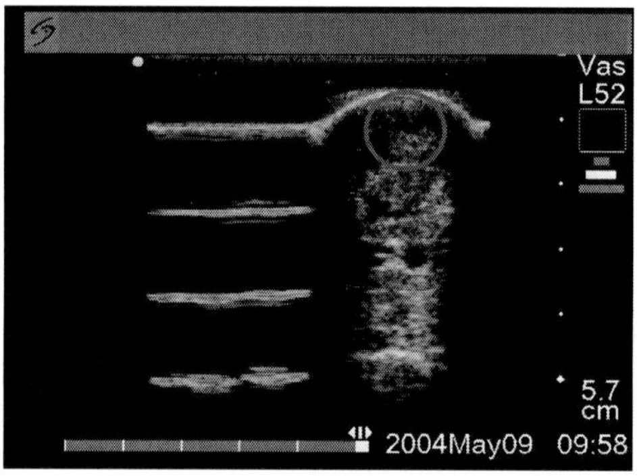

Image 5 Short axis ultrasonography of the SDFT 5-09-04.

Clinical Evaluation

Ultrasound scan imaging showed a transition to a more uniform echogenicity with a complete reduction of the SDFT core lesion to a well defined tendon cell, fibrin and collagen filled area indicative of an accelerated healing process.

At this time clinical signs showed no lameness at the walk. There was no pain elicited during flexion or palpation. There was no visible swelling or heat felt during palpation.

Third Follow-up Ultrasonographic Images

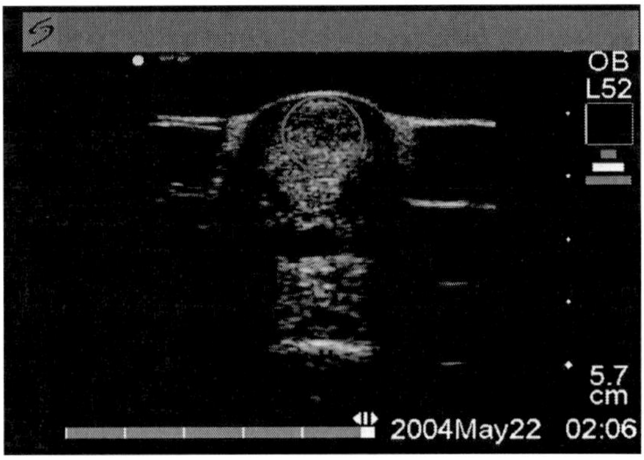

Image 6 Short axis ultrasonography of the SDFT 5-22-04.

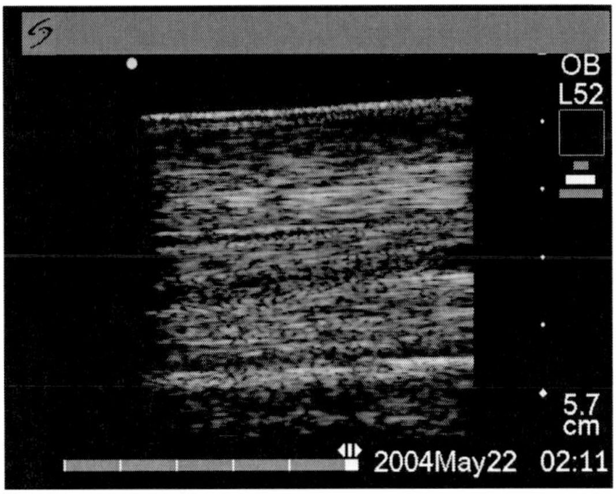

Image 7 Long axis ultrasonography of the SDFT 5-22-04.

Clinical Evaluation

Image 6. Ultrasound scan showed a uniform and normal echogenicity with a complete reduction of the SDFT core lesion to a well defined tendon cell regeneration area with no evidence of a prior SDFT lesion.

Image 7. Ultrasound scan showed proper tendon cell and collagen fiber alignment with no further evidence of a prior SDFT core lesion.

At this time clinical signs showed no lameness at the jog or gallop. There was no pain elicited during flexion or palpation. There was no visible swelling or heat felt during palpation even after the increase in physical activity in the jog or gallop. There is no more evidence of lameness to date.

Discussion

The prognosis for return to use following injury to the SDFT in athletic horses has historically been graded fair to poor. According to the literature, the healing process can take from 6 to 24 months to occur and usually terminates with further evidence of disruption of normal fiber alignment and variable loss of echogenicity at the endpoint of study per follow-up ultrasonography.

Continued evidence of disruption of normal fiber alignment and variable loss of echogenicity are usually a result of adhesions and scar tissue that occur during the course of the healing process of an injured tendon. Any disruption, malalignment or change from the normal tendon cell alignment or consistency does predispose the horse to future reinjury; due to inconsistent load force and pressure distribution on impact, in the tissue surrounding the scared-in previous injury [Crevier-Denoix N et al., 1997].

The high cost in veterinary diagnostics, care, treatment and maintenance, as well as the loss in value of the athletic horse is difficult to recuperate by the horse owner rendering the promising horse athlete a very high cost to maintain without gainful return. The high morbidity rate in the tendon injured horse, coupled with the continued threat of reinjury, translates into a substantial financial loss to the horse owner over time.

Conclusions

In this exemplary case study, we have found a rapid rate of tendon healing as well as tendon healing with the ultrasonographic evidence of the return to normal and homogeneous tendon cell integrity as proven by the uniform and normal echogenicity.

The duration of healing time, as evidenced by ultrasonography in this case study, as well as the clinical signs of healing and rapid return to function was 43 days; from the first diagnostic ultrasound scan on 4-09-04 to the last ultrasound scan on 5-22-04 to date.

This clearly shows that Equine Cyma Bioresonance treatments have surpassed any standard veterinary treatment of the SDFT injured horse by approximately 5 to 22.5 months ahead of any known standard treatments provided in the literature.

The high quality of the homogeneous, healed, tendon tissue, per ultrasonographic evidence, translated into a low to null continued morbidity or threat of reinjury.

The cost savings in veterinary diagnostics, care, treatment and maintenance over time, as well as the preservation in value of this athletic horse will further insure this horse owner a gainful return on their substantial financial investment.

Evaluation

These case studies demonstrated a significant improvement in pain and swelling, as well as objectively measurable functional improvements. Objective markers such as clinical improvement in ROM, function and return to work with no signs of lameness at the jog or the gallop.

With the encouraging results of these case studies, further investigation of the efficacy of this acoustic device is warranted.

References

A.H.J. Fleming, "*Electromagnetic self-field theory and its application to the hydrogen atom*", Physics Essays, accepted for publication May 2005. http://www.unifiedphysics.com

A.H.J. Fleming and E.B. Colorio, "*Prediction of a photonic chemistry*", 26th annual meeting of BEMS, Washington, June 20–24, 2004a. http://www.unifiedphysics.com

A.H.J. Fleming and E.B. Colorio, "*The spectroscopy of the EM field*", 3rd International Workshop on Biological effects of EMFs, Kos, 4–8 October 2004b.

A.H.J. Fleming and E.B. Colorio "*The Photon and its energy*", Biophotonics Research Institute; 2003. http://www.biophotonicsresearchinstitute.com

N. Crevier-Denoix et al. *Mechanical properties of pathological equine superficial digital flexor tendons.* Equine Vet J Suppl 1997 May; (23):23–6.

A.E. Goodship, *The pathophysiology of flexor tendon injury in the hor*se. Equine Vet Educ 1993;5:23–29.

Further information on SFT may be obtained at http://www.unifiedphysics.com

Further information on the Cyma device may be obtained at http://www. cymatherapy.com

(*Source*: E.B. Bauer, K.E. Cooper, P. Jenks, and A.H.J. Fleming, *The Effects of Acoustic Frequencies on Core Tendon Lesions of the Thoroughbred Racehorse*, Biophotonics Research Institute, Malden, MA, http://elizabethbauerconsults.com/pdfs/Tendon-Lesions-Racehorse.pdf.)

Appendix A

Frequently Asked Questions

This series of frequently asked questions (FAQs) is based on the appendix of the same name found within the mathematical physics book *Self-Field Theory: A New Mathematical Description of Physics* by Tony Fleming published by Pan Stanford Publishing in 2011. In this appendix we examine how self-field theory (SFT) changes mathematical physics and also biophysics and medicine. SFT is little short of a revolution in mathematical physics as it evolved during the 20th century. Over this period mathematical physics centred round quantum theory. Quantum theory was not able to be applied within biology; rather classical electromagnetics (EM) was used and was applicable at macroscopic domains. Yet an exact solution remained hidden and undiscovered in the mathematics. . . until now. This leads to a simplification in the way we view physics and how we can use the mathematics to predict the physics. This also has dramatic predictive implications for biology and medicine. Understanding this revolution in mathematical physics needs at least a working knowledge of SFT. A look at the index of any textbook on biology or medicine reveals that there are almost invariably no mathematical topics listed. We will not go into the reasons for this, apart from stating it as a historical fact; biologists and physicians have not had a detailed mathematical education. Now they must understand what the revolution in mathematical physics means for biology and medicine.

Vocations such as saving lives via surgery, treating illnesses via pharmaceuticals or saving ecosystems and species for posterity leave little room for mathematical research. Possibly each profession needs to know how to *manage* knowledge rather than to extend it. Where the physicist wields a syringe the biologist works with in vitro tubes or plates. Certainly neither needs to know how to wade through the formulae found within mathematical textbooks. This might seem appropriate where knowledge is a known commodity and research might reveal nothing. But science is in flux, like the period in the 19th century after Maxwell's equations were discovered and formulated from the 1860s into the 1890s. Science is changing, and the opportunities for chemistry, biology and medicine are about to be greatly augmented by SFT with its simple message of closed-form solutions to the Maxwell–Lorentz (ML) equations. It was Galen, the Roman surgeon and physician, who was willing to '*look outside the box*' by reading about all manner of learned subjects and who gave us an early example of the polymath.

According to Wikipedia:

In mathematics, an expression is said to be a **closed-form expression** if it can be expressed analytically in terms of a bounded number of certain 'well-known' functions. Typically, these well-known functions are defined to be elementary functions—constants, one variable x, elementary operations of arithmetic ($+ - \times \div$), nth roots, exponent and logarithm (which thus also include trigonometric functions and inverse trigonometric functions). By contrast, infinite series, integrals, limits, and infinite continued fractions are not permitted. Indeed, by the Stone–Weierstrass theorem, any continuous function on the unit interval can be expressed as a limit of polynomials, so any class of functions containing the polynomials and closed under limits will necessarily include all continuous functions.

SFT provides a variety of analytic expressions across physics and biophysics. It supersedes quantum theory, which is essentially a series of numerical (non-analytic) solutions across physics. While quantum theory could not be utilised within biophysics, SFT can. Up till recent times it was commonly thought within science that many problems such as atomic physics did not have closed-form solutions, also known as analytic solutions. This turns out to be false. Many solutions do have analytic solutions, including atomic physics. This induces a profound change across physics, biophysics and medicine. The impact of these solutions on biology and medicine can be seen via some recent history.

In the 20th century there were a number of biology and medical researchers who all enjoyed a period of acceptance followed by total rejection at the hands of the prevailing mainstream scientific opinion of the day. This began in the 1920s and included the following eminent scientists:

(1) Alexander Gurwitsch of Russia, who discovered that cells emit tiny numbers of biophotons as the cell cycle proceeds. Further he was the first to suggest that there was a 'division frequency', an EM frequency at which cellular replication occurs.
(2) A German émigré to the U.S., Royal Raymond Rife, in the 1920s and 1930s, investigated first the 'dark field' microscope and then how cells could react or resonate with exogenous EM sources.
(3) Ross Adey, an Australian, in the 1970s, found 'window effects' by exposing chick embryo cells to various extremely-low-frequency (ELF) and modulated ELF signals, measuring that the efflux of calcium ions, Ca^{2+}, across the cell membrane varied as a type of on-off window across the frequency spectrum as the frequency and magnitude of the exposure signal was varied.
(4) In 1987 French immunologist Jacques Benveniste published a paper that was ultimately rejected by the journal *Nature* before he, as the others mentioned above, died.

These four scientists were rejected by mainstream science, and each died after a final rejection. For a start, SFT with its new mathematics and physics and its analytic solutions is critically important in understanding the biophysical effects associated with the above experiments and their therapeutic relevance.

SFT entered the scientific lexicon and mainstream of peer review a few years ago, just after the turn of the new millennium. Although it is still going through this process of validation SFT has a much better heritage as a mathematical physics than its much older companion mathematical theory quantum field theory (QFT). This appendix puts together a list of questions that have been posed either during presentations, over the Internet on physics and biophysics forums or similar discussion groups. This particular FAQ has been written around biology and its associated disciplines. To the present, biology has had to depend on classical electromagnetics (CEM) with its limitations at the atomic domain and below. It is clear that those who are more biophysicists or physicians than mathematical biophysicists may not recognise SFT as the breakthrough that it indeed represents. Likewise many physical theoreticians are trained almost exclusively on concepts of Lagrangians, gauge theory, symmetry and a grab-bag of specialised subjects associated with the current versions of quantum theory. Without a backward glance, scant thought is given to the fact or even the possibility that quantum theory might one day become obsolete. Distinct from the ML equations based on the Maxwellian first-order equations, quantum theory is based on the second-order wave equations and the integral equation techniques used to solve them probabilistically.

While there is nothing whatsoever wrong with integral equation methods, what is at issue is an over-constrained model of the photon and its close companion, the biophoton, as a point source. This introduces a numerical error known as uncertainty and the present-day probabilistic nature of quantum theory. In addition the higher the order of the mathematics, the more complex it becomes. This applies especially to differential forms of equations. So the question boils down to whether the EM wave equations or the first-order ML equations are the basis of the mathematics. For instance, many queries concern either gauge or symmetry. There is no such thing as a gauge theory in SFT because of the simplistic way gauge applies within SFT. Again symmetry is achieved in a different way in SFT compared with QFT. Hence the issue of a non-zero-mass photon is also answered in a mathematically different fashion. Finally relativity is inherent in the bi-spinorial form of mathematics and provides a completely novel way of incorporating relativity within quantum theory. This also changes the way we view relativity to much more physically palatable. Relativity's warping of is intuitively easy to digest when it is realised that time includes the hidden time and distances of internal photon motions.

There is an enormous amount of validation yet to be performed relating to SFT. This work needs to be 'delegated' to other than those who are aware of the predictive power of SFT, similar to the case in the late 1920s and early 1930s with QM and the early concepts of quod erat demonstrandum (QED) when pioneers in various disciplines emerged to find new predictions and validate QFT in the various areas of physics to which it could successfully be applied.

Within this textbook the statement *'according to SFT'* is frequently used. This usually implies a hypothesis that is yet to be validated by experimental evidence. However, this does indicate a physical process that can be observed to correspond to the predicted mechanisms of the theory without comprehensive quantitative validation. Photonics within the atmosphere is one example. What we see are layers of chemical reactions, each different to the others. Anyone in an aircraft in the process of either attaining cruising altitude or descending in preparation for landing will have observed such layers of photonic interaction. It is the author's role to simply present the work thus far, including both the evidence garnered to date and the many hypotheses or predictions that can be associated with SFT across the widest possible gamut of physics and biophysics. That is what a theory of mathematical physics and biophysics is—a means of predicting the physical and biophysical world. The process of science is to improve on our theories. At this point in time it is our quantum theories that were first mooted around the start of the 20th century that are being questioned and found wanting. Certainly in the fullness of time, SFT will face the same curtain call. But we probably have several decades to go yet before we come to that bridge.

The first author is primarily a mathematical physicist and biophysicist with an engineering and biological background, a lifelong polymath who has discovered an exact solution to the ML system of equations that has previously remained concealed. This is the heart of SFT, an important new analytic theory of how various physical systems of particles can remain in dynamic equilibrium without expending energy. SFT turns out to be very general, not just applying to EM particles as in the hydrogen atom. SFT provides a new description across the known limits of physical knowledge. There is overwhelming evidence that SFT is a more detailed, more fundamental model of physics than QFT. In the past we have spoken of a unification of physics, and that is precisely what SFT represents.

It brings all four known forces under the one umbrella, EM strong and weak nuclear and gravitational. It predicts new gravitational forms at the galactic and higher levels of cosmological structure.

The first author has travelled widely around the globe over the past decade in an effort to spread knowledge of SFT as widely as possible. Occasionally he is asked about the evidence for SFT's predictions. While the evidence remains qualitative in the main, there is some astounding evidence that will surprise theoretical physicists who have yet to learn of SFT. To experts in a particular field born and bred on the many intricacies of QFT it may appear somewhat presumptuous as the many predictive assertions of SFT are outlined. That is as it should be. The scientific method is sceptical of new theories. Nothing is assumed without both a theoretical mechanism and a quantitative validation. The author's own bioelectromagnetic colleagues are nothing if not sceptics when it comes to new bioeffect mechanisms. For decades several bioelectromagnetic phenomena have been reviewed as un-substantiated because there has been no theoretical mechanism—'good experimental evidence, but we do not understand how such a bioeffect happens'—a form of catch 22 that Capt. Yossarian knew all too well. Now we have such a mathematical mechanism that may be behind many possible biological phenomena, and we need to complete the work of validation by matching theory with experiment. This is all work to be done within but one very important area of expertise. Speaking metaphorically, we are at the entrance of a mathematical tunnel leading to a new scientific and physical world, a 'wormhole' leading to the future. It is like the layers of a knowledge '*onion*'; the hole leads to a new level of existence.

It is hoped that the presentation in this textbook of the biological mathematics of SFT is convincing in its own right. The author has always sought to understand the mathematics at a physical level. Indeed SFT provides a much more physically intuitive picture of EM interactions than either CEM or quantum theory. It is hoped that this list of questions and answers will be useful to those who venture into SFT without the necessary mathematical depth and background. In favour of SFT is the mathematical level required to comprehend it being far simpler than QFT, and hence it can be understood by undergraduates and the casual reader with at least two years of undergraduate mathematics.

SFT may be unpalatable to the advanced exponent of QFT who has invested years of hard mental effort to master all the hand-waving and chicanery necessary to learn all the nuances of QFT. In comparison, SFT is like a stroll in the park with an old friend. This should not be taken to mean SFT is classical. SFT seems to indicate that all physics and biological life involve an infinite series of energy levels or dynamical systems of particles. Thus physics and biophysics are fractal, where each level is connected to a higher level, a lower level and within levels. Streams of field particles act as connectors between levels. This implies that there is an initial stream going back to the big bang and the absolute point in space-time that relates to all motion and life thereafter. SFT is a clearer mathematical model of reality than QFT.

What is SFT?

SFT is a theory of mathematical physics that appears to apply across many, if not all, known areas of physics and biophysics. It is an extension to the present theoretical physics that achieves unification across current knowledge of physics and biophysics. At its most fundamental it is a series of analytic solutions to a range of physical applications across EM, weak and strong nuclear and gravitational physics.

What experimental evidence exists for such claims?

(a) The primary evidence comes from the eigensolutions to the hydrogen atom that includes a theoretical derivation of Planck's constant $\hbar$. It should be noted that QFT is a heuristic formulation that proposed an eigenvalue system of equations based on the experimentally obtained estimate of $\hbar$, the eigensolution first determined by Planck in 1900. Hence SFT provides the theoretical eigenvalue basis for QFT.

$$Ax = \lambda x \tag{A.1}$$

$$\widehat{\mathrm{H}}\Psi(x,t) = i\hbar\frac{\partial}{\partial t}(x,t) \tag{A.2}$$

$$m_e v_o^2 = 2\hbar\omega_o \tag{A.3}$$

The three equations (A.1)–(A.3) demonstrate the situation described above. Equation (A.1) is the usual form of a general mathematical eigenvalue equation. Equation (A.2) is the Shrödinger equation where the left- and right-hand sides have been transposed left to right and right to left to demonstrate the fact that Planck's (reduced) constant $\hbar$ is inserted heuristically from the experiment in QM. $\hbar$ (Planck's reduced number) is a consequence of the formulation in Eq. (A.3) and does *not* come from experiment. If we rework SFT in the wave equations of QFT, this would result in a new formulation for QFT one that was deterministic, not probabilistic.

(b) The equations of SFT are identical with those of uncertainty (Heisenberg's uncertainty principle [HUP]), except the inequality condition is replaced by *two* equality conditions, one relating to the motions due to the electric currents and the other relating to the motions due to the magnetic currents of each particle.

$$\Delta p_x \Delta x \geq \hbar \tag{A.4}$$

$$m_e v_o^2 = 2\hbar\omega_o \tag{A.5}$$

$$m_e v_c^2 = 2\hbar\omega_c \tag{A.6}$$

It turns out that in EM applications only two particles can be connected to each other. They are connected by a stream of photons. Each domain from cosmological and above, down to photon and below, can only connect to a small discrete number of particles (two, three or higher).

(c) While Bohr's theory of the electron is still recognised as a useful theoretical model of the electron in the hydrogen atom, it does not include a theory of the electron's magnetic (H) field, nor the proton's. SFT supplies the missing theory for this simple model. In terms of SFT Bohr theory is said to be 'mono-spinorial', while SFT is bi-spinorial. SFT 'completes' the Bohr model of the electron and the proton. The analytic solutions obtained from SFT apply to all modes of the hydrogen atom.

(d) The bi-spinorial basis of SFT can be seen to be a form of relativity where a 'hidden variable', the photon's internal motion, accounts for some of the motion treated as external distance and time in relativity. This internal distance and time causes the dilations observed by Einstein.

Where is SFT derived from?

SFT is an exact or analytic solution that comes from the existing theory of partial differential equations. It is important to note that this analytic solution can be applied to all states of the hydrogen atom and is capable of being applied to *all* atomic and molecular structures. This may seem ludicrous or amazing to present-day chemists and scientists who use numerical methods to solve molecular structures. However, there is no mathematical reason why such a situation cannot be true. In fact SFT is capable of obtaining analytic solutions to nuclear physics. Again this is a hypothesis, and the work is yet to be performed. What can be said is that the author has obtained an analytic solution to the hydrogen diatom, commonly known as hydrogen gas, presented briefly in Chapter 3.

Is SFT related to QFT?

Yes, intimately, even though it has a much shorter history. It is based on the equations of CEM that could not be successfully applied to the hydrogen atom at the turn of the 20th century. It has taken another century to discover what was wrong with CEM and to solve the hydrogen atom analytically. Briefly, SFT is based on the first-order ML system of equations, while QFT is based on the derived second-order system of equations. Current QFT uses a different model of the photon to SFT. The photon in QFT is mono-spinorial, while it is bi-spinorial in SFT.

Why is SFT not a Lagrangian formulation?

The Maxwellian basis is much easier to work with in the same way that the finite-difference method of numerical solutions is far easier to implement than the finite-element method. It is more closely linked at an intuitive level to the important role the photon plays in physics and biophysics and to the geometric concepts seen in nature of rotation, helix, spiral, etc.

What role do gauge and symmetry play in SFT?

Gauge theory is almost trivial in SFT, and as such there is little need to consider gauge apart from recognising that each spinor rotation is O(2) and non-unitary. When we consider two particles we need to consider two field particles. In the atom this boils down to the electron, the proton and *two* photons. When we consider three particles we need to consider three field particles. In the nucleus this is three quarks (e.g., two up: up, down; one down: down, up) and three gluons.

What does SFT mean for EM? (What is wrong with CEM?)

The usual method of obtaining relativistic fields is via the Liénard Wiechert scalar and vector potentials. In other words, the fields are determined by taking the distance between two charges. In SFT this single distance is replaced by *two* distances taken between the centres of the bi-spinorial rotations of the photon. In other words there is not a single distance but two distances taken in measuring the fields. This is discussed in Chapters 1 and 3.

What does SFT mean for quantum theory in its various forms?

We need to differentiate between current implementations of QM and post-SFT implementations. If we change the way the photon is modelled inside quantum theory it will become a deterministic formulation. It may well be able to solve certain problems more accurately than SFT. Like their associated numerical methods, SFT and the modified quantum methods, there will situations where an integral equation method will be better suited to cover homogeneous regions and other applications where an inhomogeneous region will be better suited to SFT. It makes sense to retain the current nomenclature, that is, quantum theory, rather than some new term with which workers in the field have to become accustomed. Any modifications, including a different shape function for the photon; implicit incorporation of relativity via the bi-spinors; selection of mutual particles, not single particles; and deterministic output, should be seen overall as an organic process within the current terminology. That said quantum theory will change comprehensively, though not totally. It may in time be seen to be worthwhile to rename quantum mathematical methods of analysis. Perhaps integral equation methods of SFT might be called integral SFT, and the current SFT that acts on the Maxwellian and its modified forms might be called differential SFT. Whether the scalar and vector potentials or the electric (E) and H fields should be used as variables needs to be considered carefully to match a particular application. The largest change will be the use of analytic solutions in comparison with the present-day numerical solutions.

What does SFT mean for string theory?

Although SFT has other forms of forces it features a differential form of electromagnetics as a basis for gravity. The number of variables (dimensions) of space-time can be expressed as the three spatial variables and time or four. We can term these by the following indices: x, y, z and t. Then we have the differentials. If we assume that time and distance never couple, then the following differentials also apply: xx, yy, zz, xy, xz, yz and tt. This makes 11 dimensions in total. This agrees with string theory. Hence there is an intimate relationship between SFT and string theory. This gives another rationale for the validity of SFT. The current problem with string theory is its lack of application to 'everyday' or terrestrial physics.

Since the basis of SFT is the point particle we might easily change any point particle to become a string or toroid of mass m and geometry (a toroid can be specified via two radii, r_1 and r_2), thus transforming SFT into the first string theory to cover the gamut of physics as with the point-particle version of SFT. This modelling manoeuvre overcomes the main problem with string theory, its lack of application to 'everyday' physics. Whether or not matter behaves *totally* as strings is a matter for determination by modelling various situations, including the hydrogen atom. The most appropriate answer may be a combination of strings *and* point particles.

String theory can be seen to have originated from two underlying motives. First, it provides a way to achieve the wave-particle duality, and secondly it overcomes the singularity at charge points within quantum theory, a numerical error; replacing a point with a small section of string achieves the required numerical fix within quantum theory. The problem with string theory is that it has been anchored to quantum theory with its CEM basis, namely, its over-constrained model of the photon and also therefore its uncertainty principle. Hence, like quantum theory string theory can yield the standard model of particle physics and is seen to be the way energy behaves at high-energy levels. So, although it is an attractive method for mathematicians within particle physics to get their teeth into, it has to date remained somewhat removed from the everyday world of physics and remains within the ambit of quantum theory. It is not the author's desire to put hundreds and thousands of his colleagues within mathematical physics out of work. On the contrary, SFT can be seen to encompass both points *and* strings. The way photons act in their role as binding energy within atoms, molecules and all the way up to large complexes of atoms as gravitational systems is fundamentally like 'beads on a string'. The same applies to gluons. Hence if we make the changes suggested above to quantum methods, the same changes in general apply to current methods of string theory. Then having made these modifications, string theory becomes a deterministic method involving closed-form solutions across physics. Most importantly string theory might then be applied within SFT or the modified quantum theory suggested above. At present only simple paths can be analysed within the current early forms of SFT, but we can foresee new methods of analysis encompassing both string theory and SFT or string theory and the modified quantum theory. In this way string theory can be used within the wide gamut of physics to analyse the photon in its myriad roles as binding energy within atomic arrays. String theory thus becomes indispensable within future analytic methods to predict how atoms, molecules and gravitational systems behave down to the photonic level.

At another level, if string theory can be used to replace photons, particles of light, then strings can be used to approximate a series of space-time points defining the motions of strings of photons or phonons. Thus they can be useful to simplify the mathematics, possibly within the nucleus. One essential difference between SFT and string theory is that SFT is fractal; it sees fields beneath the photon structure. Note that we have no reason to believe that the photon is *not* a particle, or an infinite series of particles that look like a single particle, and is instead a string. In fact we have experimental evidence beginning with Einstein's photoelectric effect that it is particulate. Nevertheless string theory may be useful in simplifying the fractal mathematics by concluding the infinite series in pieces of string, tying up loose ends so to speak. These differences become modelling or numerical issues.

What does SFT mean for cosmology and general relativity?

We have already seen that the internal and external motions of the photon help explain in a physically intuitive way how space is not actually warped but the vision our eyes sees at relativistic speeds is warped. As Einstein knew, seeing is not always believing; straight lines could be curvilinear. In regard to the general theory of relativity, Einstein's general relativity (GR) assumes a single form of gravitation acting across the entire universe. To an approximation this is true, but the actual situation is otherwise. SFT implies three main modifications to cosmological models based on GR as it currently stands:

(1) Like CEM and its failure early in the 20th century to solve the atom, and quantum theory's lack of magnetic currents, there is a lack of any stable solution due to the failure of models to examine mutual effects between masses.

(2) The universe contains more than one type of gravitation: a tri-spinorial form applies to galaxies; a tetra-spinorial form may apply to super clusters; and maybe another form, perhaps a penta-spinor, applies to the universe itself. The overall structure of the universe is therefore not homogeneous or isotropic as assumed by GR.

(3) Another important modification relates to the photon's non-zero mass and composite structure, as described in Chapter 3. If the big bang was hot enough there would have been an initial period where a sea of sub-photonic particles existed. This may be responsible for the inflationary period when the universe expanded to near its present size at superluminal speeds. Sub-photonic particles of non-zero mass could travel at superluminal speeds and help solve the so-called horizon problem in a more intuitive way. Similarly the anisotropy observed within the universe can be explained without recourse to quantum foam theories. These modifications all have implications for the various GR solutions obtained by Friedmann, Lemaître, De Sitter, Guth and others, including Einstein's own solution obtained in 1915. Instead of the fluid dynamics of current GR models a particle-field model can give another perspective on cosmological processes. Overall this suggests an early inflationary period that finished before an evolution towards the critical condition on density, leading to a dynamic equilibrium within the universe. All effects are supported by the available cosmological evidence.

Historically, Einstein and Heisenberg crossed swords about GR and quantum theory.

A conversation between Einstein and Heisenberg:

Heisenberg: One cannot observe the electron orbits inside the atom. [...] but since it is reasonable to consider only those quantities in a theory that can be measured, it seemed natural to me to introduce them only as entities, as representatives of electron orbits, so to speak.

Einstein: But you don't seriously believe that only observable quantities should be considered in a physical theory?

'I thought this was the very idea that your relativity theory is based on?,' Heisenberg asked in surprise.

'Perhaps I used this kind of reasoning,' replied Einstein, 'but it is nonsense nevertheless. [. . .] In reality the opposite is true: only the theory decides what can be observed.'

(translated from *Der Teil und das Ganze* by W. Heisenberg)

http://www.thebigview.com/spacetime/uncertainty.html

There are theoretical similarities between GR and quantum theory. Among the most fundamental are that both are based on second-order wave equations and their associated potential theories and gauge considerations. In comparison, SFT is based on the first-order Maxwellian with its field variables that have a much reduced emphasis on gauge. Both GR and QFT are based around single-particle analyses rather than the mutual effects that couple particles together studied in SFT. Finally both GR and quantum theory employ a metric in the view of SFT to accommodate the over-constraint of the basic equations. In both cases this is linked to a theoretical requirement for a zero-mass photon.

Thus both quantum theory and GR depend upon a zero-mass photon and hence from the point of view of SFT both quantum theory and GR are theoretical approximations. For quantum theory zero mass springs from the earliest observations of beta decay and again when a negligible rest mass of the photon could hardly be compared with the seemingly endless radiation from within the nucleus of the bombs dropped on Hiroshima and Nagasaki in 1945. The cosmological principle that had its genesis in the Vatican's unscientific and dogmatic dealings with Galileo was a way to avoid having any universal centre of gravity, thus making the same mistake again. Nevertheless it is only an approximation in the light of SFT, where it is seen that non-homogeneity and anisotropy are both present in the gravitational structure itself where space is divided into different gravitational regions. This structure depends on the composite nature and non-zero mass of the photon. The space within the universe cannot be thought of as the surface of an expanding balloon other than as a theoretical approximation that holds for GR. It is known that at smaller than cosmological domains the cosmological principle does not hold, for instance, for any possible surviving location of the big bang. We may think of a biological tissue such as liver where the dielectric constant is averaged over the microstructure such as biological cells. While such an approximation is useful for numerical estimation it cannot be assumed to hold in any fine detail across smaller domains; this holds for both a homogenous isotropic model of liver and of the cosmos. This is not to denigrate GR in any way but to reveal the approximate nature of GR theory that Einstein had proposed in 1915, when it was and remains today a monumental intellectual achievement of the highest level within mathematical physics.

What is a biophoton and how does it differ from a photon?

At present we surmise that to a first approximation both the photon and the biophoton are identical. However, while a biophoton is thought to be the same as a photon, it may turn out to be essentially different. Thus far in our scientific history particle physics has studied the high end of the mass or energy spectrum. Now we have an estimate for the mass of the photon, and we can also propose a putative internal structure for the photon. The biophoton may have the same structure or maybe not. Particle physics now has a new task that applies to the low end of the spectrum. So what is different between the photon and the biophoton? We have no idea. All we know is that life is essentially different to an automaton that moves and uses a neural chip for artificial intelligence. This is something like the difference between an artificial plant that might be used for presentation and a real plant or between the two-dimensional (2D) or three-dimensional (3D) image on a television screen and real living beings. Maybe you see no difference in these cases; maybe like the authors you do. It turns out to be quite difficult to specify exactly what the differences are between living and inert systems; but when you see it, feel it, taste it and sense it, you intuitively know the difference.

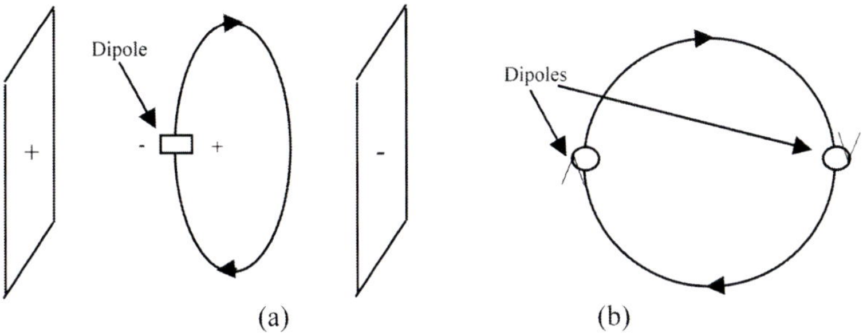

Figure A.1 (a) Dipole rotates exposed to EM by plate applicators, similar to Adey's window experiments (monopole–dipole); (b) dipoles gravitate about each other (dipole–dipole). Dipoles in cases (a) and (b) are in different orthogonal planes to the orbit; case (a) being a monopole–dipole interaction explains the orthogonal direction of the dipole compared to case (b); in both cases the orbit can be varied via the EM frequency or its magnitude.

What does SFT mean for biophysics and medicine?

Due to its magnitude it is difficult to summarise briefly what the answer is to this question. Let us first state that it is only SFT that sees exposure of the cell as related to the phenomenon of gravitation between two atomic masses. In the cases of two single atoms interacting, for example, two hydrogen atoms, the forces are a dipole–dipole interaction and the result is a hydrogen molecule. In Adey's 'window' experiments there is a dipolar protein within the membrane and an exposure applicator that provides a simple E field, a 'monopole' in this instance.[a]

However, in addition, there are many other implications for biophysics. Perhaps the most importance of these others is the existence of a photon chemistry that acts in a similar method of organisation to the well-known atomic chemistry. It may be that this is the main reason behind any healing powers of homoeopathy. The fields within solution might provide important structural effects that modulate how the aqueous dilutions might work.

[a]The dipole–dipole interaction talked about in Fig. A.1a,b (same as Fig. 4.20a,b) is an effect between the differential of an E field and another differential of an E field. These dipoles are consistent with the concept of dipoles in electrical engineering. The 'monopole' referred to in Adey's window experiments is an interaction between an electric field and a differential E field.

What does SFT mean for neuroscience?

SFT is a new mathematical description of physics and biophysics using ML equations. Rather than the probabilistic solutions of quantum theory, SFT provides closed-form expressions for physical and biophysical applications and introduces a new photonic layer of physical and biophysical interaction below the atom. This provides two new quantum numbers, in addition to the four known to quantum theory. In brief, SFT provides a new mathematical physics replacing both general relativity and quantum theory. Cellular dynamics can be described via SFT, where the deoxyribonucleic acid (DNA) has been observed to emit biogenic ultraviolet (UV) biophotons. This is based on electric and magnetic cooperation between cells within a colony of cells. A modified system of ML equations can be used to solve within the strong nuclear region of the atom, involving the photon, the phonon and the gluon. This picture of the photon, phonon and gluon leads to a new view of how DNA can store and recall sensory information as short- and long-range memories. Fractal (sleeping and waking) modes of consciousness match the internal fractal structures of the photon, phonon and gluon. All this is part of the first-known mathematical description of neuroscience, providing a theoretical framework on which to base experimental protocols.

Appendix B

The Search for a General Physical Mathematics

Prior to self-field theory (SFT), biology, bioelectromagnetics, cell biology and many other life sciences depended for theory on classical electromagnetics (CEM), with its limitations at the atomic domain and below. Since before 1927, when quantum theory was introduced, there has been an ongoing search for a general deterministic mathematics applicable across macroscopic and microscopic domains as well as the atomic and nuclear domains where probabilistic quantum methods have been preferred methods of choice. As mentioned in Chapter 1 Einstein rejected the probabilistic basis of quantum theory and searched unsuccessfully for a deterministic and unifying field theory that might be applicable across physics.

Another leading figure in this search was Herbert Fröhlich, who along with Einstein, can be considered the scientific forebear of all who have sought the path of a unifying mathematics. Fröhlich was involved in many areas of science, including particle physics, with contributions to nuclear forces and meson theory, and biology, including dielectric theory and coherence. The first author is grateful to Dr Peter Rowlands of the Oliver Lodge Laboratory, Department of Physics at the University of Liverpool for a copy of the autobiographical book *Herbert Fröhlich FRS: A Physicist Ahead of His Time*. Fröhlich was Chair of Theoretical Physics at the University of Liverpool from 1948 till retiring in 1973 (from the position of Chair but not from further scientific endeavours) until his death in 1991, aged 85. Fröhlich had used quantum theory in the area of semiconductors to great effect, writing his first book *Elektronentheorie der Metalle* (Springer, Berlin) in 1936. Like many other German and Eastern European scientists of this era he fled persecution. He escaped St Petersburg (then Leningrad) and the Stalinist purges of 1935, being interned as an 'alien' in London in 1940. Many of his ideas, including biological coherence, have been taken up by others, including the first author whose work on diffusion within the membrane had its genesis in Fröhlich's ideas encountered during his PhD. Of particular interest to SFT is Fröhlich's 'bilocal' extension of Dirac theory to take into account the magnetic currents in quantum theory. This is equivalent to the complete Bohr theory solution of SFT that also takes into account the magnetic currents and motions of the electron. Thus Fröhlich and Hertz, whose potentials in application to the half-wave dipole antenna were the original stimulus for the bi-spinorial basis within SFT, can be closely associated with the origins of SFT.

Appendix C

Self-Field Theory: How Widespread Is Life within the Galaxy?

The question whether life is found throughout our galaxy was first posed via the Drake equation, which seeks the number of technologically advanced civilizations capable of sending intelligent radio signals. A related but far less restrictive equation can be proposed that seeks to estimate the number of planets in our galaxy that have at least once in their evolved history supported some level of biological life form. The basis of this equation is the recent finding that biodiversity on Earth acts as a resonance phenomenon involving galactic cycles. Using the mathematics of self-field theory (SFT) a new tri-spinorial dynamics due to the gravitational tides of galaxies was proposed; hence biodiversity found within the fossil record of marine species can be explained via galactic dynamics. A link between cosmological evolution and biological evolution was also proposed, whereby cosmological expansion corresponds to the size and structure of biological species on Earth as they have evolved. While biodiversity is observably cyclic, life itself is broadband, hence the myriad, fractal adaptations and mutations that have been observed over the past four billion years on planet Earth. Given the expanding proto-solar system, there is every possibility that the other planets may have also been able to seed life early in the galactic evolution; each planet at some stage received similar energies to that on proto-Earth. Whether life survived as the solar system evolved is unknown. Europa, the moon of Jupiter, is a location where the seeding of life may have been able to survive. Possible microfossils on Martian meteorites show life may have been seeded and survived elsewhere (Fig. C.1). Spectroscopy at radio frequencies in interstellar space in 1994 revealed the presence of the simplest amino acid, glycine (NH_2CH_2COOH). Since biodiversity is broadband, life may be widespread in elemental form, at least within our galaxy.

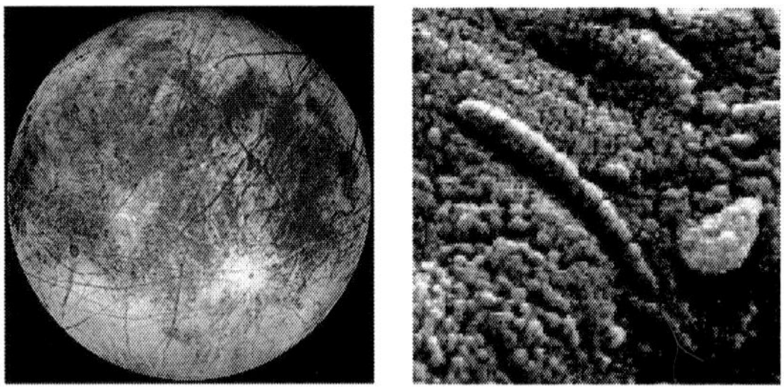

The elongated structure in the center maybe a microfossil

Figure C.1 (a) Europa, the moon of Jupiter and (b) possible microfossil from Mars.

References

Fleming, A.H.J., *Self-Field Theory: Biodiversity may be a Resonance Process*, PIERS 2010 Cambridge, MA, 2010.

Fleming, A.H.J., *Self-Field Theory: A Possible Gravitational Structure for Galaxies*, PIERS 2010 Cambridge, MA, 2010.

Fleming, A.H.J., *Self-Field Theory: Cosmological and Biological Evolution May Be Linked*, PIERS 2011 Marrakesh, MO, 2010.

Fleming, A.H.J., *Self-Field Theory: A New Mathematical Description of Physics*, Pan Stanford Publishing, 2011.

Appendix D

Self-Field Theory: A Biophotonic Model of Cellular Replication

This paper presents a model for cellular replication based on the biophysics of self-field theory (SFT) with its photon chemistry, a recently discovered level of organisation below atomic chemistry resulting from the non-zero mass of the photon. Like bubbles that form into foam, cells occur in large populations known as colonies that cooperate amongst themselves to achieve replication. Overall (1) the cells interact via cell–cell signalling of electric (E) and magnetic (H) fields (termed 'biophotons'), and (2) the chromosomes emit biophotons of specific frequencies within the ultraviolet (UV) range related to the continuously changing molecular state of the chromosomes as replication proceeds. The various states of the chromosomes are controlled by their photonic states. The states result from cell–cell interactions of first electric and then magnetic fields arising from the diffusion of dipolar proteins within the plasma membranes of various sectors of the colony. These two processes couple, inducing metaphase and anaphase. Dipolar proteins diffuse within the membrane of cells in response to intra- or extracellular endogenous or exogenous E and H fields. These diffusive processes involve two feedback loops, (1) electrical feedback between membrane proteins of a fertilised cell and those of surrounding cells in a north–south orientation and (2) magnetic feedback between biophotons emitted by chromosomes and proteins diffusing within the plasma membranes of neighbouring cells in an east–west orientation. Experimental observations, including Gurwitsch's work in the 1920s–1930s, cytoplasmic 'stirring' just before metaphase and dielectric theory within the extremely-low-frequency (ELF) range support the hypothesis. The UV frequencies inherent in the photonic states of the deoxyribonucleic acid (DNA) may have first originated from energy cycles such as the seasons available within the proto-solar system, while the ELF signals are biogenic in origin. Life may have first been a way of incorporating cyclic energy changes within a small body of tissue. In this early period, cooperation to establish life rather than competition for available food may have been a modus operandi within unicellular species. Cooperation was necessary since the available energy was dropping as cosmic expansion continued. Biophotonics research continues within areas of growth and development, repair and maintenance, genetics and evolution.

References

Fleming, A.H.J., Electromagnetic self-field theory and its application to the hydrogen atom, *Phys. Essays*, **18**(3): 265–285, 2005.

Fleming, A.H.J., and E.B. Bauer, *A Predicted Photon Chemistry*, BEMS-26, Washington DC, 2004.

Fleming, A.H.J., *Self-Field Theory*: *Analytic Estimate for the Mass of the Photon*, PIERS 2009, Moscow, Russia, 2009.

Fleming, A.H.J., *Self-Field Theory*: *A New Mathematical Description of Physics*, Pan Stanford Publishing, 2011.

Fleming, A.H.J., and E.B. Bauer, *Inside the Photon*: *A Journey to Health*, Pan Stanford Publishing, to be published.

Fleming, A.H.J., *Self-Field Theory*: *Biodiversity May Be a Resonance Process*, PIERS 2010 Cambridge, MA, 2010.

References

Many of the following texts have been chosen because they are *not* arcane academic texts gathering dust in university libraries and are fairly readily available. In this age of the Internet, book stores such as Amazon, Barnes and Noble, and Abe Books provide a repository of original texts or digitized data (e-books) from around the globe. The scientific pool of knowledge contained within them is thus widespread and not esoteric or restricted in nature. It has been the authors' pleasure to read these texts or parts thereof in order to gain an overview of the world's scientific and medical knowledge at this point in time at the start of the 21st century. Many of the references in the diffusion and computational section are taken from the first author's PhD.

General History of Science

Barnett J.E., *Time's Pendulum: The Quest to Capture Time, from Sundials to Atomic Clocks*, Plenum, Cambridge, 1998.

Bodanis D., *$E=mc^2$: A Biography of the World's Most Famous Equation*, Macmillan, London, UK, 2000.

Braun J., *Music in Ancient Israel/Palestine: Archaeological, Written, and Comparative Sources*, Eerdmans, 2002.

Clegg B., *Light Years*, Piatkus, London, UK, 2001.

Cropper W.H., *Great Physicists*, Oxford University Press, New York, NY, 2001.

Ferris T., *The Red Limit: The Search for the Edge of the Universe*, Perennial, New York, NY, 2002.

Gibbon J., *Science a History 1543–2001*, Penguin, London, UK, 2002.

Gleick J., *Genius: Richard Feynman and Modern Physics*, Abacus, London, UK, 2000.

Hamilton J.D., *Faraday: The Life*, Harper Collins, London, UK, 2003.

Hawking S., *On the Shoulders of Giants: The Great Works of Physics and Astronomy*, Running Press, Philadelphia, PA, 2002.

Hunt V.H., *Origins in Acoustics: The Science of Sound from Antiquity to the Age of Newton*, Yale University Press, New Haven, CT, 1978.

Rhodes R., *The Making of the Atomic Bomb*, Simon and Schuster, 1988.

Rouse Ball W.W., *A Short Account of the History of Mathematics*, 4th ed., Dover Publications, New York, NY, 2003.

Thorne K.S., *Black Holes and Time Warps: Einstein's Outrageous Legacy*, W.W. Norton, New York, NY, 1995.

Whittaker E., *A History of the Theories of Aether and Electricity*, Vol. 1, Philosophical Library, New York, NY, 1954.

Whittaker E., *A History of the Theories of Aether and Electricity*, Vol. 2, Harper & Brother, New York, NY, 1960.

Williams, H.S., *The Story of Nineteenth Century Science*, Harper & Brothers, 1904.

Yaghjian A., *Relativistic Dynamics of a Charged Sphere: Updating the Lorentz-Abraham Model*, 2nd ed., Lecture Notes in Physics, Springer, 2005.

Mathematics

Jancewicz B., *Multivectors and Clifford Algebra in Electrodynamics*, World Scientific, Singapore, 1988.

Misra J.C., *Biomathematics: Modeling and Simulation*, World Scientific, Singapore, 2006.

Schutz B.F., *Geometrical Methods of Mathematical Physics*, Cambridge University Press, Cambridge, 1980.

Thomas G.B., Jr., *Calculus and Analytic Geometry*, 3rd ed., World Student Series, Addison-Wesley, 1966.

Electromagnetics

Barut A.O., *Electrodynamics and Classical Theory of Fields and Particles*, Dover, New York, NY, 1980.

Harrington R.F., *Time Harmonic Electromagnetic Fields*, Mc-Graw-Hill, New York, NY, 1961.

Kraus J.D., *Antennas*, McGraw-Hill Electrical and Electronic Engineering Series, McGraw-Hill, 1950.

Lehnert B., and S. Roy, *Extended Electromagnetic Theory: Space-Charge in Vacuo and the Rest Mass of the Photon*, World Scientific Series in Contemporary Chemical Physics, World Scientific, 1999.

Melia F., *Electrodynamics, Chicago Lectures in Physics*, University of Chicago Press, London, UK, 2001.

Jackson J.D., *Classical Electrodynamics*, 3rd ed., John Wiley & Sons, New York, NY, 1999.

Von Hippel A., *Dielectrics and Waves*, John Wiley & Sons, New York, NY, 1962.

Maxwell J.C., *Treatise on Electricity and Magnetism*, 3rd ed., Vols. 1 and 2, Dover Publications, 1954.

Schwartz M., *Principles of Electrodynamics*, Dover, New York, NY, 1987.

Quantum Theory

Condon E.U., and H. Odabasi, *Atomic Structure*, Cambridge University Press, Cambridge, 1980.

Heisenberg W., *The Physical Principles of the Quantum Theory*, Dover, New York, NY, 1949.

Heitler W., *The Quantum Theory of Radiation*, Dover, New York, NY, 1984.

House J.E., *Fundamentals of Quantum Mechanics*, Academic Press London, UK, 1998.

McMahon D., *Quantum Field Theory Demystified*, McGraw Hill, 2008.

Miller A.I., *Early Quantum Electrodynamics: A Sourcebook*, Cambridge University Press, Cambridge, 1995.

Srednicki M., *Quantum Field Theory*, Cambridge University Press, 2007.

Physics

Yariv A., *Quantum Electronics*, 3rd ed., Wiley & Sons, 1988.

N.W. Ashcroft and N.D. Mermin, *Solid State Physics*, Holt-Saunders International Editions, 1976.

Einstein A., *Relativity, the Special and General Theory*, Three Rivers Press, New York, NY, 1961.

Einstein A., *The Meaning of Relativity, Fifth Edition: Including the Relativistic Theory of the Non-Symmetric Field*, MJF Books, New York, NY, 1984.

Einstein A., Relativity, the special and general theory, in *Albert Einstein in His Own Words*, Portland House, New York, NY, 2000.

Fowler C.M.R., *The Solid Earth: An Introduction to Global Geophysics*, Cambridge University Press, 1990.

Hawking S., *A Brief History of Time*, Bantam, London, UK, 1988.

Henson R., *The Rough Guide to Weather*, Rough Guides, 2002.

Jenny H., *Cymatics: A Study of Wave Phenomena and Vibration*, 3rd ed., Macromedia Press, July 2001.

Kittel C., *Quantum Theory of Solids*, 2nd ed., Wiley, 1987.

Maxwell J.C., *Matter and Motion*, Great Minds Series, Prometheus Books, reprint edition, 2002.

Pauling L., *The Nature of the Chemical Bond and the Structure of Molecules and Crystals; An Introduction to Modern Structural Chemistry*, 3rd ed., George Fisher Baker Nonresident Lecture, Cornell University Press, 1960.

Pierce A.D., *Acoustics: An Introduction to Its Physical Principles and Applications*, American Institute of Physics, 1989.

Stevens K.W.H., *Magnetic Ions in Crystals*, Princeton University Press, 1997.

Rindler W., *Relativity: Special, General, and Cosmological*, Oxford University Press, Oxford, 2004.

Nuclear and Particle Physics

Cottingham W.N., and D.A. Greenwood, *An Introduction to the Standard Model of Particle Physics*, Cambridge University Press, Cambridge, UK, 1998.

Heisenberg W., *Introduction to the Unified Field Theory of Elementary Particles*, Interscience, New York, NY, 1966.

Krane K.S., *Introductory Nuclear Physics*, John Wiley & Sons, New York, NY, 1988.

Perkins D.H., *Introduction to High Energy Physics*, 4th ed., Cambridge University Press, Cambridge, UK, 2003.

Watson A., *The Quantum Quark*, Cambridge University Press, 2004.

Astronomy and Cosmology

Adams F., and G. Laughlin, *The Five Ages of the Universe: Inside the Physics of Eternity*, Touchstone, New York, NY, 1999.

Binney J., and S. Tremaine, *Galactic Dynamics*, Princeton Series in Astrophysics, 1987.

Klacka J., *Galactic Tide*, arXiv-0912.3112v1 [astro-ph.GA] Dec. 16, 2009.

Mather J.C., and J. Boslough, *The Very First Light*, Penguin Books, London, UK, 1998.

Medvedev M.V., and A.L. Melott, Do extragalactic cosmic rays induce cycles in fossil diversity? *Astrophys. J.*, **664**:879–889, 2007.

North J.D., *The Measure of the Universe*, Clarendon Press, Oxford, 1960.

Rohde R.A., and R.A. Muller, Cycles in fossil diversity, *Nature*, **434**:208–210, 2004.

Biology and Bioelectromagnetics

Ahmed A.J., and A. Alizad, *Biomedical Applications of Vibration and Acoustics in Therapy, Bioeffect and Modeling*, ASME, New York, NY, 2008.

Beloussov L.V., V.L. Voeikov, and V.S. Martynyuk, *Biophotonics and Coherent Systems in Biology*, Springer, 2006.

Binhi V.N., *Magnetobiology: Underlying Physical Problems*, Academic Press, 2002.

Chang R., *Physical Chemistry with Applications to Biological Systems*, 2nd ed., Prentice Hall College, 1977.

Gagliardi L.J., Electrostatic force in prometaphase, metaphase, and anaphase-A chromosome motions, *Endogeneous Physical Fields in Biology*, URSI International Symposium July 1–3, Prague, Czech Republic, 2002.

Gilbert F.G., *Developmental Biology*, 9th ed., Sinauer Associates, 2010.

Haeger K., *The Illustrated History of Surgery: Revised and Updated by Sir Roy Calne FRS*, Harold Starke, London, UK, 2000.

Harrison R., The outgrowth of the nerve fiber as a mode of protoplasmic movement, *J. Exp. Zool.*, **9**:787–846, 1910

Ho M.-W., F.A. Popp., and U. Warnke, *Bioelectrodynamics and Biocommunication*, World Scientific, 1994.

Ling G.N., *Life at the Cell and Below-Cell Level*, Pacific Press, Melville, NY, 2001.

Polk C., *CRC Handbook of Biological Effects of Electromagnetic Fields*, CRC Press, 1986.

Nakamura J., and M. Hiramatsu, Ultra-weak photon emission from human hand: influence of temperature and oxygen concentration on emission, *J. Photochem. Photobiol. B: Biol.*, **80**:156–160, 2005.

Popp, F.A., Q. Gu, and K.H. Li, Biophoton emission: experimental background and theoretical approaches, *Mod. Phys. Lett.*, **B8**:1269–1296, 1994.

Popp F.A., *About the Coherence of Biophotons, Microscopic Quantum Coherence*, Proceedings of an International Conference, World Scientific, River Edge, NJ, 1999.

Rosch P.J., and M.S. Markov, *Bioelectromagnetic Medicine*, Marcel Dekker, December 2004.

Lund E.J., *Bioelectric Fields and Growth*, University of Texas, 1947.

Zewail A. (Ed.), *Physical Biology: From Atoms to Medicine*, Imperial College Press, 2008.

Self-Field Theory

Fleming A.H.J., *Self-Field Theory: A New Mathematical Description of Physics*, International Scientific Congress 2010, Fundamental Problems of Natural Sciences and Engineering, St. Petersburg, Russia, July 26–30, 2010.

Fleming A.H.J., *Self-Field Theory: Are Cosmological and Biological Evolution Linked?* International Scientific Congress 2010, Fundamental Problems of Natural Sciences and Engineering, St. Petersburg, Russia, July 26–30, 2010.

Fleming A.H.J., *Self-Field Theory*: *Biodiversity May Be a Resonance Process*, PIERS-10, Cambridge, MA, June 2010.

Fleming A.H.J., *Self-Field Theory*: *A Possible Gravitational Structure for Galaxies*, PIERS-10, Cambridge, MA, June 2010.

Fleming A.H.J., *Self-Field Theory*: *Biophotons and EPR*, PIERS-10, Cambridge, MA, June 2010.

Fleming A.H.J., *Self-Field Theory: New Photonic Insights*, PIERS-10, Xi'an, China, March 2010.

Fleming A.H.J., *Analytic Estimate of the Mass of the Photon*, PIERS-09, Moscow, Russia, September 2009.

Fleming A.H.J., *The Role of Endogenous and Exogenous E-Fields in Metaphase*, PIERS-09, Beijing, China, March 2009.

Fleming A.H.J., *Self-Field Theory: The Spectroscopy of the Photon*, BEMS-30, San Diego, June 2008.

Fleming A.H.J., *Self-Field Theory: Analytic Spectroscopy of the Ordinary Photon*, 2nd International Conference on Electromagnetic Fields, Health and Environment, Wroclaw, Poland, September 10–12, 2007.

Fleming A.H.J., E.B. Bauer, R. Bergeron, M. Dahle, and J. Enge, *Acoustic/ Magnetic Field Assisted Perfusion*, 2nd International Conference on Electromagnetic Fields, Health and Environment Wroclaw, Poland, September 10–12, 2007.

Fleming A.H.J., *Self-Field Theory: The Mass of the Photon and the Fine Structure Constant*, BEMS-29, Kanazawa, Japan, June 2007.

Fleming A.H.J., *Self-Field Theory: Mathematical Perspective*, BEMS-28, Cancun, Mexico, June 2006.

Fleming A.H.J., *Electromagnetic Self-Field Theory and Its Application to the Hydrogen Atom*, Vol. 18, 3, Physics Essays, 2005.

Fleming A.H.J., and E.B. Bauer, *Case Study: Part A Electromagnetic Self-Field Theory*, BEMS-27, Dublin, June 2005.

Bauer E.B., K. Cooper, P. Jenks, and A.H.J. Fleming, *Case Study: Part B the Effects of Acoustic Frequencies on Core Tendon Lesions of the Thoroughbred Racehorse*, BEMS-27, Dublin, June 2005.

Fleming A.H.J., and E.B. Colorio (nee Bauer), *The Spectroscopy of the EM Field: A Predicted Photon Chemistry*, 3rd International Workshop on Biological Effects of EM Fields, Kos, Greece, October 2004.

Fleming A.H.J., and E.B. Colorio (nee Bauer), *A Predicted Photon Chemistry*, BEMS-26, Washington DC, June 2004.

Other References

Fleming A.H.J., *Towards Computational Methods for Studying Cellular Effects due to EM Field*, Applied Computational Electromagnetics Conference, Naval Postgraduate College, Monterey, California, March 15–19, 1999.

Fleming A.H.J., and P.M. Farrell, *A Model for Ionic Diffusion in a Liquid due to Static and Time-Varying E- and B-Fields*, Sixteenth Annual Meeting Bioelectromagnetics Society, Copenhagen, June 12–17, 1994.

Fleming A.H.J., and K.H. Joyner, Estimates of absorption of radiofrequency radiation by the embryo and fetus during pregnancy, *Health Phys.*, **63**:149–159, 1992.

Fleming A.H.J., and K.H. Joyner (Eds.), Special issue on bioelectromagnetic computations, *Appl. Comput. Electromagn. Soc. J.*, **7**(2), 1992.

Fleming A.H.J., and K.H. Joyner, *Radio-Frequency Radiation Exposure and Pregnancy*, Joint Annual Conference Australian Radiation Protection Society and Australasian College of Physical Scientists and Engineers in Medicine, Adelaide, September 24–28, 1990.

Fleming A.H.J., A finite element method for composite scatterers, in Morgan M.A. (Ed.), *Finite Element and Finite Difference Methods in Electromagnetic Scattering, Progress in Electromagnetics Research*, Vol. 2, 69–112, Elsevier, New York, 1990.

Hocking B., K.H. Joyner, and A.H.J. Fleming, Implanted medical devices in workers exposed to radio-frequency radiation, *Scand. J. Work Environ. Health*, **17**:1–6, 1990.

Fleming A.H.J., and K.H. Joyner, Moment method analysis of radiation and scattering by thin wires in an infinite conducting medium, *Spec. Issue ACES J. Electromagn. Code Valid.*, 51–74, 1989.

Diffusion and Computational References

Abromowitz M., and I.A. Stegun, *Handbook of Mathematical Functions*, Dover Publications, New York, 1970.

Adair R.K., Constraints on biological effects of weak extremely low frequency electromagnetic fields, *Phys. Rev. A*, **43**(2):1039–1048, 1991a.

Adair R.K., Biological effects on the cellular level of electric field pulses, *Health Phys.*, **61**(3):395–399, 1991b.

Adair R.K., Effects of ELF magnetite fields on biological magnitite, *Bioelectromagnetics*, **13**:1–4, 1993.

Adey W.R., Frequency and power windowing in tissue interactions with weak electromagnetic fields, special issue on biological effects and medical applications of electromagnetic energy, *IEEE Proc.*, **68**(1):119–125, 1980.

Adey W.R., issue interactions with nonionizing electromagnetic fields, *Phys. Rev.*, **61**(2):435–514, 1981.

Adey W.R., and A.R. Sheppard, Cell surface ionic phenomena in transmembrane signaling to intracellular enzyme systems, in Blank M., and E.

Findl (Eds.), *Mechanistic Approaches to Interactions of Electric and Electromagnetic Fields with Living Systems*, 365–388, Plenum Press, New York, 1987.

Allan M.P., and D.J. Tildesley, *Computer Simulation of Liquids*, Clarendon Press, Oxford, 1987.

Amaro M.L., L.B. Bhuiyan, and C.W. Outhwaite, Micellar solutions with added salt a Monte Carlo and modified Poisson-Boltzmann study, *Mol. Phys.*, **86**, 725–735, 1995.

American National Standards Institute, *American National Standard Safety Levels with Respect to Human Exposure to Radio-Frequency Electromagnetic Fields, 3 kHz to 300 GHz*, American National Standards Institute, New York, 1991.

Anders R., NAC-3, ACES Software Library Number #011, 1987.

Anderson V., and K.H. Joyner, Specific absorption rate levels measured in a phantom head exposed to radio frequency transmissions from analog hand-held phones, *Bioelectromagnetics*, **15**, 1995.

Arkin H., L.X. Xu, and K.R. Holmes, Recent developments in modelling heat transfer in blood perfused tissues, *IEEE Trans.*, **BME-41**(2):97–107, 1994.

Ashcroft N.W., and N.D. Mermin, *Solid State Physics*, Holt-Saunders, 1976.

Astrahan, M.A., G. Luxton, M.D. Sapozink, and Z. Petrovich, The accuracy of temperature measurement from within an interstitial microwave antenna, *Int. J. Hyperthermia*, **4**(6):593–607, 1988.

Barber, P.W., Electromagnetic power deposition in prolate spheroid models of man and animals at resonance, *IEEE Trans.*, **BME-24**:513–521, 1977a.

Barber, P.W., Resonance electromagnetic absorption by nonspherical dielectric objects, *IEEE Trans.*, **MTT-25**:373–381, 1977b.

Barnes F.S., Interaction of DC electric fields with living matter, in Polk C., and E. Postow (Eds.), *CRC Handbook of Biological Effects of Electromagnetic Fields*, 99–119, CRC Press, Florida, 1986.

Barnes F.S., and M. Seyed-Madani, Some possible limits on the minimum electrical signals of biological significance, in Blank M., and E. Findl (Eds.), *Mechanistic Approaches to Interactions of Electric and Electromagnetic Fields with Living Systems*, 339–347, Plenum Press, New York, 1987.

Bassen H., P. Herchenroeder, A. Cheung, and S. Neuder, Evaluation of an implantable electric-field probe within finite simulated tissues, *Radio Sci.*, 15–25, 1977.

Bassett C.A.L., Fundamental and practical aspects of therapeutic uses of pulsed electromagnetic fields (PEMFs), *Crit. Rev. Biomed. Eng.*, **17**(5):451–527, 1989.

Bawin S.M., L.K. Kaczmarek, and W.R. Adey, Effects of modulated VHF fields on the central nervous system, *Ann. NY Acad. Sci.*, **247**:74–91, 1975.

Bawin S.M., and W.R. Adey, Sensitivity of calcium binding in cerbral tissue to weak environmental electric fields oscillating at low frequency, *Proc. Natl. Acad. Sci. U S A*, **73**:1999–2003, 1976.

Bawin S.M., L.S. Kinney, D.E. House, and W.T. Joines, Multiple power-density windows and their possible origin, *Bioelectromagnetics*, **10**:115–128, 1989.

Bennet M.V.L., and S. Obara, Ionic mechanisms and pharmacology of electroreceptors, in Bullock T.M., and W. Heiligenberg (Eds.), *Electroreception*, Wiley-Interscience, New York, 1986.

Birdsall C.K., Particle-in-cell charged particle simulation, plus monte carlo collisions with neutral atoms, PIC-MCC, *IEEE Trans.*, **PS-19**(2):65–85, 1991.

Blackman C.F., L.S. Kinney, D.E. House, and W.T. Joines, Ionic processes at membrane surfaces: the role of the electrical double layers in electrically stimulated ion transport, in Blank M., and E. Findl (Eds.), *Mechanistic Approaches to Interactions of Electric and Electromagnetic Fields with Living Systems*, 1–15, Plenum Press, New York, 1987.

Blechschmidt E., *The Stages of Human Development before Birth, an Introduction to Human Embryology*, W.B. Saunders, Philadelphia, 1961.

Bo W.J., N.T. Wolfman, W.A. Krueger, and I. Meschan, *Basic Atlas of Sectional Anatomy*, W.B. Saunders, Philadelphia, 1980.

Brenner H., Coupling between the translational and rotational Brownian motions of rigid particles of arbitrary shape. I Helicoidally isotropic particles, *J. Colloid Sci.*, **20**:104–122, 1955.

Brenner H., Coupling between the translational and rotational Brownian motions of rigid particles of arbitrary shape. II General theory, *J. Colloid Interface Sci.*, **23**:407–436, 1967.

Brune D., and S. Kim, Predicting protein diffusion coefficients, *Proc. Natl. Acad. Sci. U S A*, **90**:3835–3839, 1993.

Bullock T.H., and W. Heiligenberg (Eds.), *Electroreception*, Wiley-Interscience, New York, 1986.

Burke G.J., and A.J. Poggio, *Numerical Electromagnetics Code (NEC)-Method of Moments*, Naval Ocean System Centre, Office of Naval Research Contract N00014-71-C-0187, 1981.

Bydder G.M., Interpretation and clinical application of magnetic resonance imaging, *IEEE Trans.*, **MAG-26**:2089–2091, 1990.

Carpenter D.O., and S. Ayrapetyan (Eds.), *Biological Effects of Electric and Magnetic Fields*, Academic Press, San Diego, 1994.

Carstensen E.L., *Biological Effects of Transmission Line Fields*, Elsevier, New York, 194–202, 1987.

Carson P.L., *Medical Ultrasound Fields and Exposure Measurements, Proceedings of the Twenty-Second Annual Meeting of NRCP*, 287–307, NRCP, Bethesda, MD, 1988.

Chang H., and K.K. Mei, Scattering of EM waves by buried or partly buried body of revolution, *IEEE Symp. Antennas Propag.*, 653–656, 1981.

Char B.W., K.O. Geddes, G.H. Gonnet, B.L. Leong, M.B. Monagan, and S.M. Watt, *Maple V*, Springer-Verlag, New York, 1991.

Chay T.R., and J. Keizer, Minimal model for membrane oscillations in the pancreatic β-cell, *Biophys. J.*, **42**:181–190, 1983.

Chay T.R., and J. Rinzel, Bursting, beating and chaos in an excitable membrane model, *Biophys. J.*, **47**:357–366, 1985.

Chen F.F., *Introduction to Plasma Physics*, Plenum Press, New York, 1974.

Chen K.M., H.R. Chuang, and C.J. Lin, Quantification of interaction between ELF-LF electric fields and human bodies, *IEEE Trans.*, **BME-33**(8):747–755, 1986.

Chen K.M., and B.S. Guru, Focal hyperthermia as induced by RF radiation of simulacra with embedded tumors and as induced by EM fields in a model of a human body, *Radio Sci.*, **12**(6S):27–37, 1977.

Chew W.C., *Waves and Fields in Inhomogeneous Media*, Van Nostrand Reinhold, New York, 1990.

Chiabrera A., and B. Bianco, The role of the magnetic fields in the em interaction with ligand binding, in Blank M., and E. Findl (Eds.), *Mechanistic Approaches to Interactions of Electric and Electromagnetic Fields with Living Systems*, 79–95, Plenum Press, New York, 1987.

Ciccotti G., and G. Jacucci, Direct computation of dynamical response by molecular dynamics: the mobility of a charged lennard-jones particle, *Phys. Rev. Lett.*, **35**(12):789–792, 1975.

Clarke H.T. (Ed.), *Ion Transport across Membranes*, Academic Press, New York, 1954.

Cleveland R.F., and T.W. Athey, Specific absorption rate (SAR) in models of the human head exposed to hand-held UHF portable radios, *Bioelectromagnetics*, **10**:173–186, 1989.

Constable R.T., P.B. Dunscombe, A. Tsoukatos, and K. Malaker, Perturbation of the temperature distribution in microwave-irradiated tissues due to the presence of metallic thermometers, *Med. Phys.*, **14**(3):385–388, 1987.

Cole K.S., *Membranes, Ions and Impulses, a Chapter of Classical Biophysics*, University of California Press, Berkeley, 1972.

Crane H.R., The *g* factor of the electron, *Sci. Am.*, **231**(1):72–85, 1968.

Darnell J., H. Lodish, and D. Baltimore, *Molecular Biology*, Scientific American Press, New York, 1986.

Davis W.A., D.G. Sweeney, and W.L. Stutzman, MiniNEC II: an improved version of MiniNEC for personal computers, URSI Meeting, June 15–19, 1987.

Dawson J.M., Particle simulation of plasmas, *Rev. Mod. Phys.*, **55**(2):403–447, 1983.

Debye P., *Polar Molecules*, Dover, 1929.

Dimbylow P.J., and S.M. Mann, SAR calculations in an anatomically realistic model of the head for mobile communication transceivers at 900 MHz and 1.9 GHz, *Phys. Med. Biol.*, **39**(10):1537–1553, 1994

Dunscombe P.B., J. McLellan, and K. Malaker, Heat production in microwave-irradiated thermocouples, *Med. Phys.*, **13**(4):457–461, 1986.

Durney C.H., Electromagnetic dosimetry for models of humans and animals: a review of theoretical and numerical techniques, special issue on biological effects and medical applications of electromagnetic energy, *IEEE Proc.*, **68**:33–40, 1980.

Durney C.H., M.F. Iskander, H. Massoudi, S.J. Allen, and J.C. Mitchell, *Radiofrequency Radiation Dosimetry Handbook*, 3rd ed., Brooks Air Force Base, TX, USAF School of Aerospace Medicine, SAM-TR-80-32, 1980.

Durney C.H., C.C. Johnson, P.W. Barber, H. Massoudi, M.F. Iskander, J.L. Lords, D.K. Ryser, S.J. Allen, and J.C. Mitchell, *Radiofrequency Radiation Dosimetry Handbook*, 2nd ed., Brooks Air Force Base, TX, USAF School of Aerospace Medicine, SAM-TR-78-22, 1978.

Durney, C.H., H. Massoudi, and M.F. Iskander, *Radiofrequency Radiation Dosimetry Handbook*, 4th ed., Brooks Air Force Base, TX, USAF School of Aerospace Medicine, SAM-TR-85-73, 1986.

Durney C.H., C.K. Rushforth, and A.A. Anderson, Resonant AC-DC magnetic fields: calculated response, *Bioelectromagnetics*, **9**:315–336, 1988.

Ehrenberg B., D.L. Farkas, E.N. Fluhler, Z. Lojewskz, and L.M. Loew, Membrane potential induced by external electric field pulses can be followed with a potentiometric dye, *Biophys. Soc.*, **51**:833–837, 1987.

Ekstrom P., and D. Wineland, The isolated electron, *Sci. Am.*, **243**(2):91–101, 1980.

El-Khoury G., R.A. Bergman, and W.J. Montgomery, *Sectional Anatomy by MRI/CT*, Churchill Livingstone, New York, 1990.

Einstein A., *Investigations on the Theory of Brownian Movement*, Fürth R. (Ed.), translated by A.D. Cowper, Dover, 1956.

Ellis W.V., Pain control using high-intensity pulsed magnetic stimulation, *Bioelectromagnetics*, **14**:553–556, 1994.

Encarta CD Encyclopedia, Electroconvulsive therapy, Microsoft Corporation, Funk & Wagnall's, New York, 1993.

Feynman R.P., R.B. Leighton, and M. Sands, *The Feynman Lectures on Physics*, Commemorative Issue, Addison-Wesley, 1989.

Fessard A. (Ed.), *Handbook of Sensory Physiology*, Springer-Verlag, New York, 1974.

Fishman H.M., On the responsiveness of elasmobranch fishes to weak electric fields, in Blank M., and E. Findl (Eds.), *Mechanistic Approaches to Interactions of Electric and Electromagnetic Fields with Living Systems*, 431436, Plenum Press, New York, 1987.

Fleming A.H.J., Numerical analysis of electromagnetic scattering by axisymmetric inhomogeneous composite antenna structures, *M. App. Sc. Thesis*, Department of Mathematical Sciences, Faculty of Technology, Chisholm Institute of Technology, Caulfield East, Victoria, Australia, 1987.

Fleming A.H.J., A finite element method for composite scatterers, in Morgan M.A. (Ed.), *Finite Element and Finite Difference Methods in Electromagnetic Scattering, Progress in Electromagnetic Research*, vol. 2, Elsevier, New York, 1990.

Fleming A.H.J., and P.M. Farrell, *A Model for Ionic Diffusion in a Liquid due to Static and Time-Varying E- and B-Fields*, Sixteenth Annual Meeting Bioelectromagnetics Society, Copenhagen, June 12–17, 1994.

Fleming A.H.J., and Joyner K.H. (Eds.), Special issue on bioelectromagnetic computations, *Appl. Comput. Electromagn. Soc. J.*, **7**(2), 1992.

Fleming A.H.J, V. Lubinas, and K.H. Joyner, *Calculation of Electric Fields in Tissue near Metallic Implants*, 1992 Asia-Pacific Microwave Conference, Adelaide, August 11–14, 1992.

Foster K.R., and H.P. Schwan, Dielectric properties of tissue, in Polk C., and E. Postow (Eds.), *Handbook of Biological Effects of Electromagnetic Fields*, 27–96, CRC Press, Florida, 1986.

Frankel R.B., Biological effects of static magnetic fields, in Polk C., and E. Postow (Eds.), *CRC Handbook of Biological Effects of Electromagnetic Fields*, 169–196, CRC Press, Florida, 1986a.

Frankel R.B., Magnetite and magnetotaxis in bacteria and algae, Maret G., N. Boccara, and J. Kiepenheuer (Eds.), in *Biophysical Effects of Steady Magnetic Fields*, 173–179, Proceedings of the Workshop, Les Houches, France, February 25–March 5, Springer-Verlag, New York, 1986b.

Furnell G.D., *Potential Distributions for Ellipsoidal Bodies in Uniform Fields*, Materials Research Laboratory, Defence Science and Technology Organization, Salisbury, Adelaide, 1991.

Galt S., J. Sandblom, and V. Hamnenius, Theoretical study of the resonant behaviour of an ion confined to a potential well in a combination of ac and dc magnetic fields (biomembrane transport), *Bioelectromagnetics*, **14**(4):299–314, 1993.

Gandhi O. (Ed.), Special issue on biological effects and medical applications of electromagnetic energy, *IEEE Proc.*, **68**(1):126–132, 1980.

Gandhi O.P., J.Y. Chen, and W. Ding, Electromagnetic absorption in the human head and neck for cellular telephones at 835 MHz, submitted to *IEEE MGW Lett.*, 1994.

Glatzmaier, G.A., and P.H. Roberts, A three-dimensional convective dynamo solution with rotating and finitely conducting inner core and mantle, *Phys. Earth Planetary Interiors*, **91**(1–3):63–75, 1995.

Grant G., R. Cadossi, and G. Steinberg, Protection against focal cerebral ischemia following exposure to a pulsed electromagnetic field, *Bioelectromagnetics*, **15**:205–216, 1994.

Grundler W., F. Kaiser, F. Keilmann, and J. Walleczek, Mechanisms of EM interaction with cellular systems, *Die Naturwissenschaften*, **79**(12):551–559, 1992.

Grynkiewicz G., M. Poenie, and R.Y. Tsien, A new generation of Ca^{2+} indicators with greatly improved flourescence properties, *J. Biol. Chem.*, **260**:3440–3450, 1985.

Guy A.W., *Biophysics-Energy Absorption and Distribution*, AGARD Lecture Series No. 78 on Radiation Hazards, Report No. LS-78, 1975.

Hafemeister D. Background paper on power line fields and public health, Open Letter, http://www.calpoly.edu/~dhafemei, 1996.

Hagmann, M.J., O.P. Ghandi, and C.H. Durney, Numerical calculation of electromagnetic energy deposition for a realistic model of man, *IEEE Trans.*, **MTT-27**:804–809, 1979.

Halle B., On the cyclotron resonance mechanism for magnetic field effects on transmembrane ion conductivity, *Bioelectromagnetics*, **9**:381–385, 1988.

Haus H.A., and J.R. Melcher, *Electromagnetic Fields and Energy*, Prentice-Hall, New Jersey, 1989.

Harrington R.F., *Time-Harmonic Electromagnetic Fields*, McGraw-Hill, New York, 1961.

Harrington R.F., *Field Computation by Moment Methods*, Macmillan, New York, 1968.

Hart F.X., Using a spread sheet program to model the interaction of low-frequency electric fields with inhomogeneous, irregularly shaped objects, *J. Bioelectr.*, **8**(2):201–226, 1989.

Hill S.C., and D.Q. Chowdhury, Three-dimensional SAR distributions computed in a multilayered cylindrical model for electromagnetic hyperthermia, *IEEE Trans.*, **MTT-37**:1192–1199, 1989.

Ho H.S., Energy absorption patterns in circular triple-layered cylinders exposed to planewave sources, *Health Phys.*, **31**:97–108, 1976.

Hocking B., Occupational health – a physician's viewpoint as exemplified by radiofrequency radiation concerns, in Repacholi M.H. (Ed.), *Non-Ionizing Radiations – Physical Characteristics, Biological Effects and Health Hazard Assessment, Proceedings of the International Non-Ionizing Radiation Workshop*, 422–423, Melbourne, IRPA, 1988.

Hocking B., K.H. Joyner, and A.H.J. Fleming, Implanted medical devices in workers exposed to radio-frequency radiation, *Scand. J. Work Environ. Health*, **17**:1–6, 1990.

Huang H.W., Mobility and diffusion in the plane of cell membrane, *J. Theor. Biol.*, **40**:11–17, 1973.

Hui S.W., Electric field-induced calcium flux and changes in cell shape, motility, and cytoskeleton, in Carpenter D.O., and S. Ayrapetyan (Eds.), *Biological Effects of Electric and Magnetic Fields*, Academic Press, San Diego, 1994.

Hultqvist B., and L. Stenflo, *Physics of the Hot Plasma in the Magnetosphere*, Proceedings of Nobel Symposium No. 30 held at Kiruna, Sweden, Plenum Press, 1975.

Hurt W.B., Multiterm Debye dispersion relations for permittivity of muscle, *IEEE Trans.*, **BME-32**:60–64, 1985.

Hush J.M., and R.L. Overall, Cortical microtubule reorientation in higher plants: dynamics and regulation, *J. Microsc.*, **181**(part 2):129–139, 1996.

International Labour Organization, *Radiation Protection of Workers (Ionizing Radiations),* International Labour Office, Geneva, 1987.

International Non-Ionizing Radiation Committee of the International Radiation Protection Association (IRPA), Guideline on limits of exposure to radio-frequency electromagnetic fields in the frequency range from 100 kHz to 300 GHz, *Health Phys.*, **54**(1):115–123, 1988.

International Non-Ionizing Radiation Committee of the International Radiation Protection Association (IRPA), Interim guideline on limits of exposure to 50/60 Hz electric and magnetic fields *Health Phys.*, **58**(1):113–122, 1991.

Iskander M.F., and A.M. Tumeh, Design optimization of interstitial antennas, *IEEE Trans.*, **BME-36**:238–246, 1989.

Jackson J.D., *Classical Electrodynamics*, 309–346, Wiley, 1962.

Jackson J.D., *Classical Electrodynamics*, 2nd ed., Wiley, 1975.

Jaffe L.F., Control of development by ionic currents, in Cone R.A., and J.E. Dowling (Eds.), *Membrane Transduction Mechanisms*, Raven, New York, 199–231, 1979.

Jaffe L.F., and R. Nuccitelli, Electric controls in development, *Ann. Rev. Biophys. Bioeng.*, **6**:445–476, 1977.

Johnson C.C., C.H. Durney, P.W. Barber, H. Massoudi, S.J. Allen, and J.C. Mitchell, *Radiofrequency Radiation Dosimetry Handbook*, Brooks Air Force Base, TX, USAF School of Aerospace Medicine, SAM-TR-76-35, 1976.

Joyner K.H., Extremely low frequency electromagnetic radiation and interactions at the cell membrane, *Radiat. Prot. Aust.*, **6**:118–125, 1988.

Joyner K.H., B. Hocking, A.H.J. Fleming, and I.P. Macfarlane, *Metallic Implants and Exposure to Radio-Frequency Radiation*, Seventh International Congress of the International Radiation Protection Association, Sydney, April 10–17, 1988.

Julian A.J., J.C. Logan, and J.W. Rockway, *MININEC; A Mini-Numerical Electromagnetic Code*, Naval Ocean System Centre, Technical Document 516, 1982.

Kalmijn A.J., The detection of electric fields from inanimate and animate sources other than electric organs, in Fessard A. (Ed.), *Handbook of Sensory Physiology Vol 3: Electro-Receptors and Other Specialized Receptors in Lower Vertebrates*, 147–200, Springer-Verlag, New York, 1974.

Kalmijn A.J., Biophysics of geomagnetic field detection, *IEEE Trans.*, **MAG-17**(1):1113–1123, 1981.

Karlon W.J., S.R. Eisenberg, and J.L. Lehr, Defibrillation current density distribution:a three-dimensional finite element model of the canine thorax, *IEEE EMBS Symp.*, 1991.

Kaune W.T., Macroscopic dosimetry of power-frequency electric and magnetic fields, *Bioelectromagnetics*, Suppl. **1**:11–14, 1992.

Keeton W.T., T.S. Larkin, and D.M. Windsor, Normal fluctuations in the earth's magnetic field influence pidgeon orientation, *J. Comput. Physiol.*, **95**:95–103, 1974.

King, R.W.P., *Antennas in Dissipative Media and in Subsurface Communications*, Technical Report, Division of Engineering and Applied Physics, Harvard University, 1975.

King R.W.P., Imbedded bare and insulated antennas, *Trans. IEEE*, **BME-24**:253–260, 1977.

King, R.W.P., and C.W. Harrison, Jr., *Antennas and Waves – A Modern Approach*, MIT Press, Massachusetts, 1969.

King, R.W.P., and K. Iizuka, The complete electromagnetic field of a half-wave dipole in a dissipative medium, *Trans. IEEE*, **AP-11**:275–285, 1963.

King R.W.P., and B.H. Sandler, Cylindrical antenna in a dissipative medium program, *Trans. IEEE*, **AP-21**:410–411, 1973.

King R.W.P., B.H. Sandler, and T.T. Wu, Cylindrical antennas immersed in arbitrary homogeneous isotropic media, *J. Appl. Phys.*, **40**:5049–5065, 1969.

King R.W.P., and G.S. Smith, *Antennas in Matter*, MIT Press, Cambridge, MA, 1981.

Kinouchi Y., S. Tanimoto, K. Sato, H. Yamaguchi, and H. Miyamoto, Effects of static magnetic fields on diffusion in solution, *Bioelectromagnetics*, **9**:159–166, 1988.

Kirschvink L., and M.M. Walker, Biogenic magnetite in higher organisms and current status of the hypothesis of ferrimagnetic magnetoreception, in Maret G., N. Boccara, and J. Kiepenheuer (Eds.), *Biophysical Effects of Steady Magnetic Fields*, 173–179, Proceedings of the Workshop, Les Houches, France, February 25–March 5, Springer-Verlag, New York, 1986.

Kirschvink J.L., A. Kobayashi-Kirschvink, J.C. Diaz-Ricci, and S.J. Kirschvink, Magnetite in human tissues, *Bioelectromagnetics*, Suppl. **1**:101–113, 1992.

Kittel C., *Elementary Statistical Physics*, Wiley, 1958.

Krassowska W., and J.C. Neu, Response of a single cell to an external electric field, *Biophys. J.*, **66**:1768–1776, 1994.

Kritikos K.N., and H.P. Schwan, Hot spots generated in conducting spheres by electromagnetic waves and biological implications, *IEEE Trans.*, **BME-19**:53–58, 1972.

Kuster N., Multiple multipole method applied to an exposure safety study, *ACES J.*, **7**(2):43–60, 1992.

Lednev V.V., Possible mechanism for the influence of weak magnetic fields on biological systems, *Bioelectromagnetics*, **12**:71–75, 1991.

Lee S.H., and J.C. Rasaiah, Molecular dynamics simulation of ionic mobility. 1. Alkali metal cations in water at 25°C, *J. Chem. Phys.*, **101**:6964–6974, 1994.

Liboff A.R., Cyclotron resonance in membrane transport, in Chiabrera A., C. Nicolini, and H.P. Schwan (Eds.), *Interactions between Electromagnetic Fields and Cells*, 281–296, Plenum Press, London, UK, 1985.

Liboff A.R., S.D. Smith, and B.R. Mcleod, Experimental evidence for ion cyclotron resonance of membrane transport, in Blank M., and E. Findl (Eds.), *Mechanistic Approaches to Interactions of Electric and Electromagnetic Fields with Living Systems*, 109–132, Plenum Press, New York, 1987.

Liboff A.R., and W.C. Parkinson, Search for ion-cyclotron resonance in a Na^+-transport system, *Bioelectromagnetics*, **12**:77–83, 1991.

Lin J.C., Computer methods for field intensity predictions, in Polk C., and E. Postow (Eds.), *CRC Handbook of Biological Effects of Electromagnetic Fields*, 273–314, CRC Press, Florida, 1986.

Lin J.C., A.W. Guy, and C.C. Johnson, Power deposition in a spherical model of man exposed to 1-20 MHz EM fields, *IEEE Trans.*, **MTT-21**:791–797, 1973.

Liu L.M., and S.F. Cleary, Absorbed energy distribution from radiofrequency electromagnetic radiation in a mammalian cell model: effect of membrane-bound water, *Bioelectromagnetics*, **16**:160–171, 1995.

Logan N.A., Survey of some early studies of the scattering of plane waves by a sphere, *Proc. IEEE*, **53**(8):773–785, 1965.

Luben R.A., Effects of low-energy electromagnetic fields (pulsed and DC) on membrane signal transduction processes in biological systems, *Health Phys.*, **61**(1):15–28, 1991.

Lund E.J., *Bioelectric Fields and Growth*, University of Texas Press, Austin, 1947.

MacNeal B.E., *An Introduction to MSC/EMAS*, MacNeal-Schwendler, 1990.

Mallorqui J.J., A Broquetas, L. Jofre, and A. Cardama, Non-invasive active thermometry with a microwave tomographic scanner in hyperthermia treatments, *ACES J.*, **7**(2):121–127, 1992.

Maret G., N. Boccara, and J. Kiepenheuer (Eds.), *Biophysical Effects of Steady Magnetic Fields*, Proceedings of the Workshop, Les Houches, France, February 25–March 5, Springer-Verlag, New York, 1986.

Maraner P., *Adiabatic Motion of a Quantum Particle in a Two-Dimensional Magnetic Field*, Università di Parma, Dipartimento di Fisica, UPRF-95-434, 1995.

McConnell J., *Rotational Brownian Motion and Dielectric Theory*, Academic Press, London, UK, 1980.

Mcleod B.R., and A.R. Liboff, Cyclotron resonance in cell membranes; the theory of the mechanism, in Blank M., and E. Findl (Eds.), *Mechanistic Approaches to Interactions of Electric and Electromagnetic Fields with Living Systems*, 97–108, Plenum, New York, 1987.

Melville D., F. Paul, and S. Roath, High gradient magnetic separation of red blood cells from whole blood, *IEEE Trans.*, **MAG-11**:1701–1704, 1975.

Moilel, R.A., S.K. Wolfson, Jr., R.G. Selker, and S.B.Weiner, Materials for selective tissue heating in a radiofrequency EM field for the combined chemothermal treatment of brain tumors, *J. Biomed. Mater. Res.*, **10**:327–334, 1976.

Monaghan J.J., and C.M. Shapcott, Radiation from accelerated point dipoles in circular and Keplerian orbits, *Aust. J. Phys.*, **25**:197–206, 1972.

Morgan M.A., and K.K. Mei, Finite element computation of scattering by inhomogeneous bodies of revolution, *IEEE Trans.*, **AP-27**:203–214, 1979.

Morgan M.A., Finite element calculation of microwave absorption by the cranial structure, *IEEE Trans.*, **BME-28**:687–695, 1981.

Morgan, M.A., C.H. Chen, S.C. Hill, and P.W. Barber, Finite element – boundary integral formulation for electromagnetic scattering, *Wave Motion, North Holland*, **6**:91–103, 1984.

Morgan M.A., Private correspondence, 1989.

Morgan M.A., Coupled potentials for electromagnetic fields in inhomogeneous media, in Morgan M.A. (Ed.), *Finite Element and Finite Difference Methods in Electromagnetic Scattering, Progress in Electromagnetic Research*, vol. 2, Elsevier, New York, 1990.

Moriyama E., N. Matsumi, T. Shiraishi, T. Tamiya, T. Satoh, K. Matsumoto, T. Furuta, and A. Nishimoto, Hyperthermia for brain tumors: improved delivery with a new cooling system, *Neurosurgery*, **23**(2):189–195, 1989.

MacNeal, B.E. (Ed.), *MSC/EMAS Users Manual*, MacNeal-Schwendler, Milwaukee, 1994.

Murray R.W. The function of the ampullae of Lorezini of elasmobranchs, in Cahn P. (Ed.), *Lateral Line Detectors*, 277–293, Indiana University Press, Indiana, 1967.

National Council on Radiation Protection (US), *Radiofrequency Electromagnetic Fields: Properties, Quantities and Units, Biophysical Interaction, and Measurement*, NCRP Report No. 67, NCRP, Washington, 1981.

National Radiation Protection Board (UK), *Electromagnetic Fields and the Risk of Cancer*, Report of an advisory group on nonionizing radiation, NRPB, Chilton, 1992.

Neuder S.M., *Electromagnetic Fields in Biological Media, Part II – The SCAT Program, Multilayered Spheres, Theory and Applications*, US Deptartment of Health, Education and Welfare, Bureau of Radiological Health, HEW Publication (FDA) 79-8072, Rockville, MD, 1979.

Nuccitelli R. (Ed.), *Ionic Currents in Development, Progress in Clinical and Biological Research*, vol. 210, Liss, 1986.

Obara S., and M.V.L. Bennett, Mode of operation of ampullae of Lorenzini of the skate raja, *J. Gen. Phys.*, **60**:534–577, 1972.

Papoulis A., *Probability, Random Variables, and Stochastic Processes*, 515–553, International student edition, McGraw-Hill, 1965.

Paulsen K.D., X. Jia, and D.R. Lynch, 3D bioelectromagnetic computation on finite elements, *ACES J.*, **7**(2):9–25, 1992.

Penman S., Rethinking cell structure, *Proc. Natl. Acad. Sci. U S A*, **92**:5251–5257, 1995.

Pethig R., Electrical properties of biological tissue, in *Modern Bioelectricity*, 125–179, Marcel Dekker, New York, 1988.

Physical Science Study Committee, *Physics*, Heath, Boston, 1965.

Plonsey R., and R.C. Barr, *Bioelectricity – A Quantitative Approach*, Plenum Press, New York, 1988.

Polk C., and E. Postow (Eds.), *CRC Handbook of Biological Effects of Electromagnetic Fields*, 1–24, CRC Press, Florida, 1986.

Polk C., Modification of charge distribution at boundaries between electrically dissilimar media, in Blank M., and E. Findl (Eds.), *Mechanistic Approaches to Interactions of Electric and Electromagnetic Fields with Living Systems*, 63–77, Plenum Press, New York, 1987.

Poo M., In situ electrophoresus of membrane components, *Ann. Rev. Biophys. Bioeng.*, **10**:245–276, 1981.

Presti D., and J.D. Pettigrew, Ferromagnetic coupling to muscle receptors as a basis for geomagnetic field sensitivity in animals, *Nature*, **285**:99–101, 1980.

Rafferty C.N., R.O. Phillips, and A.W. Guy, Dosimetry Workshop: extremely low frequency electric and magnetic fields, *Bioelectromagnetics*, Suppl. **1**, 1992.

Ramo S., J.R. Whinnery, and T. Van Duzer, *Fields and Waves in Communication Electronics*, 460–461, Wiley, New York, 1965.

Ranjan, R., and N.V. Thakor, Electrical stimulation of cardiac myocytes, *Ann. Biomed. Eng.*, **23**(6):812–821, 1995.

Rasmussen H., The calcium messenger system, parts 1 and 2, *New Engl. J. Med.*, 1094–1101, 1164–1170, 1986.

Richmond J.H., Rradiation and scattering by thin-wire structures in a homogeneous conducting medium, computer program description, *Trans. IEEE*, **AP-23**:365, 1975.

Reif F., *Fundamentals of Statistical and Thermal Physics*, 461–493, McGraw-Hill, New York, 1965.

Reif F., *Statistical Physics*, vol. 5, Berkeley Physics Course, McGraw-Hill, New York, 1967.

Rosenweig R.E., *Ferrohydrodynamics*, Cambridge University Press, Cambridge, 1985.

Rosenweig R.E., Negative viscosity in a magnetic fluid, *Science*, **271**:614–615, 1996.

Royer G.M., A Monte Carlo procedure for potential theory problems, *IEEE Trans.*, **MTT19**:813–818, 1971.

Sadiku M.N.O., and D.T. Hunt, Solution of Dirichlet problems by the Exodus method, *IEEE Trans.*, **MTT-40**:89–95, 1992.

Sathiaseelan V., A. Taflove, M.J. Piket-May, C. Reuter, and B.B. Mittal, Application of numerical modelling techniques in electromagnetic hyperthermia, *ACES J.*, **7**(2):61–71, 1992.

Sala F., and A. Hernández-Cruz, Calcium diffusion modelling in a spherical neuron, *Biophys. J.*, **57**:313–324, 1990.

Scaife B.K.P., *Complex Permittivity*, Clarendon Press, Oxford, 1989.

Schams H., and J. Bretscher, *Ultrasonographic Diagnosis in Obstetrics and Gynecology*, Springer-Verlag, New York, 1975.

Schelkunoff S.A., *Antennas: Theory and Practice*, 218–223, Wiley, New York, 1952.

Schrödinger E., *Wave Mechanics*, Blackie, London, UK, 1928.

Schwan H.P., Electrical properties of tissues and cells, in Lawrence J.H., and A. Tobias (Eds.), *Advances in Biological and Medical Physics*, 147–209, Academic Press, New York, 1957.

Schwartz J.L, D.E. House, and G.A.R. Mealing, Exposure of frog hearts to CW or amplitude-modulated VHF fields:selective efflux of calcium ions at 16 Hz, *Bioelectromagnetics*, **11**:349–358, 1990.

Selden E.S., and G.J. Bourke, *Antenna Modelling Program*, MBA Information Systems, Office of Naval Research Contract N00014-72-C-0187, 1974.

Semm P., T. Schneider, and L. von Irath, Effects of an earth-strength magnetic field on electrical activity of pineal cells, *Nature*, **288**:607–608, 1980.

Shapiro A.R., R.F. Lutomirski, and H.T. Yura, Induced fields and heating within a cranial structure irradiated by an EM plane wave, *IEEE Trans.*, **MTT-19**:187–196, 1971.

Sinnott D.H., Private correspondence, 1985.

Smith C.W., and Best S., *Electromagnetic Man, Health and Hazard in the Electromagnetic Environment*, Dent, London, UK, 1990.

Sowers A.E., and C.R. Hackenbrock, Rate of lateral diffusion of intramembrane particles: measurement by electrophoretic displacement and rerandomization, *Proc. Natl. Acad. Sci. U S A*, **78**:6246–6250, 1981.

Standards Association of Australia, *Radio-Frequency Radiation Part 1: Maximum Exposure Levels – 100 kHz to 300 GHz*, Standards House, North Sydney, AS2772.1, 1990.

Störmer C, *The Polar Aurora*, Oxford, 1955.

Sullivan D.M., O.P. Gandhi, and A. Taflove, Use of finite difference time-domain method for calculating EM absorption in man models, *IEEE Trans.*, **BME-35**:179–186, 1988.

Taylor H.C., and R.W.M. Lau, Evaluation of clinical hyperthermia treatment using time domain finite difference modelling technique, *ACES J.*, **7**(2):85–96, 1992.

Tenforde T.S., Dosimetry, biological effects and interaction mechanisms of extremely-low-frequency fields, *Radiat. Protect. Aust.*, **9**(3):85–93, 1991.

Tenforde T.S., Microscopic dosimetry of extremely-low-frequency electric and magnetic fields, *Bioelectromagnetics*, Suppl. **1**:61–66, 1992.

Thompson C.J., K.M. Briggs, P.M. Farrell, A.H.J. Fleming, B. Hocking, K. Joyner, V. Anderson, and A.W. Wood, Nonlinear dynamics of charged particles with combined AC-DC electromagnetic fields, *Phys. A*, **220**:471–484, 1995.

Uhlenbeck G.E., and S. Goudsmit, Spinning electrons and the structure of spectra, *Nature*, **117**:264–265, 1926.

Van De Hulst H.C., *Light Scattering by Small Particles*, Wiley, 1957.

Von Hippel A., *Dielectrics and Waves*, Wiley, 1962.

Voitenko Y.M., Anomalous magnetic diffusion in coronal current layers, *Solar Phys.*, **161**:197–200, 1995.

Walker M.M., M.E. Bitterman, and J.L. Kirschvink, Experimental and correlational studies of responses to magnetic field stimuli by different species, in Maret G., N. Boccara, and J. Kiepenheuer (Eds.), *Biophysical Effects of Steady Magnetic Fields*, 173–179, Proceedings of the Workshop, Les Houches, France, February 25–March 5, Springer-Verlag, New York, 1986.

White R.G., G.J. Hyde, and R.L. Overall, Microtubule arrays in regenerating Mougeotia protoplasts may be oriented by electric fields, *Protoplasma*, **158**:73–85, 1990.

Wolke S., U. Neibig, R. Elsner, F. Gollnick, and R. Meyer, *A Setup for Measurement of Intracellular Calcium during High Frequency Application*, Sixteenth Annual Meeting Bioelectromagnetics Society, Copenhagen, June 12–17, 1994.

Wood A.W., V. Lubinas, K.H. Joyner, and B.A. Hocking, Calcium efflux from toad heart: a replication study, in Blank M. (Ed.), *Electricity and Magnetism in Biology and Medicine*, 484–484, San Francisco Press, San Francisco, 1993.

World Health Organization, *Health Criteria Document 69: Magnetic Fields*, World Health Organization, Geneva, 1987.

Zwamborn A.P.M., P.M. van den Berg, J. Mooibroek, and F.T.C. Koenis, Computation of three-dimensional electromagnetic field distributions in a human body using the weak form of the CGFFT method, *ACES J.*, **7**(2):26–42, 1992.

Index